Contemporary Biographies in Chemistry

Contemporary Biographies in Chemistry

SALEM PRESS
A Division of EBSCO Publishing
Ipswich, Massachusetts

GREY HOUSE PUBLISHING

Cover Photo: Peggy Whitson, American biochemist and astronaut;
© Yuri Kochetkov/epa/Corbis.

Copyright © 2013, by Salem Press, A Division of EBSCO Publishing, Inc. All rights reserved. No part of this work may be used or reproduced in any manner whatsoever or transmitted in any form or by any means, electronic or mechanical, including photocopy, recording, or any information storage and retrieval system, without written permission from the copyright owner. For permissions requests, contact proprietarypublishing@ebscohost.com.

Contemporary Biographies in Chemistry, 2013, published by Grey House Publishing, Inc., Amenia, NY, under exclusive license from EBSCO Publishing, Inc.

∞ The paper used in these volumes conforms to the American National Standard for Permanence of Paper for Printed Library Materials, Z39.48-1992 (R1997).

Library of Congress Cataloging-in-Publication Data

Contemporary biographies in chemistry.
  pages cm
  Includes bibliographical references and indexes.
  ISBN 978-1-58765-997-3
  1. Chemists--Biography. 2. Women chemists--Biography.
  QD21.C66 2013
  540.92'2--dc23

2012044526

ebook ISBN: 978-1-58765-999-7

PRINTED IN THE UNITED STATES OF AMERICA

Contents

Publisher's Note	vii	Steitz, Joan A.	174
Contributors List	ix	Tanaka, Koichi	182
		Terrett, Nicholas	187
Barton, Jacqueline K.	3	Wambugu, Florence	192
Belcher, Angela	12	Wells, Spencer	199
Bertozzi, Carolyn R.	21	Whitson, Peggy	207
Catlin, Don H.	27	Wüthrich, Kurt	212
Chauvin, Yves	37	Zewail, Ahmed H.	218
Chizmadzhev, Yuri Aleksandrovich	43	**Appendixes**	**223**
Ciechanover, Aaron	47	Historical Biographies	225
Djerassi, Carl	56	Amedeo Avogadro	225
Doudna, Jennifer	63	Joseph Black	231
Good, Mary L.	69	Robert Boyle	236
Gupta, Mahabir P.	74	Henry Cavendish	242
Harris, Eva	78	John Dalton	248
Heath, James R.	86	Josiah Willard Gibbs	254
Hershko, Avram	94	Antoine-Laurent Lavoisier	260
Jones, Monty	101	Dmitry Ivanovich Mendeleyev	265
Kenyon, Cynthia	107	Linus Pauling	274
Kiessling, Laura	118	Robert Burns Woodward	281
Moncada, Salvador	122	Bibliography	289
Olopade, Olufunmilayo	130	Selected Works	295
Paabo, Svante	138		
Rubin, Edward M.	146	**Indexes**	**297**
Solomon, Susan	157	Geographical Index	299
Stefansson, Kari	164	Name Index	301

Publisher's Note

Contemporary Biographies in Chemistry is a collection of thirty-one biographical sketches of "living leaders" in the fields of physics. All of these articles come from the pages of *Current Biography*, the monthly magazine renowned for its unfailing accuracy, insightful selection, and the wide scope of influence of its subjects. These up-to-date profiles draw from a variety of sources and are an invaluable resource for researchers, teachers, students, and librarians. Students will gain a better understanding of the educational development and career pathways—the rigors and rewards of a life in science—of the contemporary scientist to better prepare themselves for a scientific career.

The geographical scope of *Contemporary Biographies in Chemistry* is broad; selections span the Eastern and Western Hemispheres, covering numerous major geographical and cultural regions. While most of the figures profiled are practicing scientists in their respective fields, the selection also includes scientifically trained government officials, institutional directors, and other policy leaders who are helping to shape the future of science by setting agendas and advancing research. Scientific fields covered range from biochemistry to genetics, to pharmacology and oncology, and from chemical engineering to molecular biology.

Articles in *Contemporary Biographies in Chemistry* range in length from roughly 1,000 to 4,000 words and follow a standard format. All articles begin with ready-reference listings that include birth details and concise identifications. The articles then generally divide into several parts, including the Early Life and Education section, which provides facts about the scientists' early lives and the environments in which they were reared, as well as their educational background; and Life's Work, a core section that provides straightforward accounts of the periods in which the profiled subjects made their most significant contributions to science. Often, a final section, Significance, provides an overview of the scientists' places in history and their contemporary importance. Essays are supplemented by bibliographies, which provide starting points for further research.

As with other Salem Press biographical reference works, these articles combine breadth of coverage with a format that offers users quick access to the particular information needed. For convenience of reference, articles are arranged alphabetically by scientists' names, and an appendix lists scientists' names by their country of origin. In addition, a general bibliography offers a comprehensive list of works for students seeking out more information on a particular scientist or subject, while a separate bibliography of selected works highlights the significant published works of the scientists profiled. An appendix consisting of ten historical biographies of "Great Chemists," culled from the Salem Press *Great Lives* series, introduces readers to scientists of historical significance integral to the genesis of chemistry and whose work and research revolutionized science.

The editors of Salem Press wish to extend their appreciation to all those involved in the development and production of this work; without their expert contribution, projects of this nature would not be possible. A full list of contributors appears at the beginning of this volume.

Contributors List

Mike Batistick
Matt Broadus
Damitri Cavalli
Andrew I. Cavin
In-Young Chang
Forrest Cole
Matthew Creamer
Christopher Cullen
Jennifer Curry
Kathleen A. D'Angelo
Sara J. Donnelly
Karen E. Duda
William Dvorak
Ronald Eniclerico
Karen E. Euda
Kaitlen J. Exum
Dan Firrincili
Terence J. Fitzgerald
Willie Gin
Peter G. Herman
Josha Hill
Martha A. Hostetter
Louisa Jennings
Virginia Kay
Patrick Kelly
David J. Kim
Dmitry Kiper
Christopher Luna
Nicholas W. Malinowski
Christopher Mari
Paul B. McCaffrey

Margaret R. Mead
David Moldawer
Majid Mozaffari
Bertha Muteba
Tracy O'Neill
Geoff Orens
Joanna Padovano
Yongsoo Park
Kenneth J. Partridge
Jamie E. Peck
Constantina Petropoulos
David Ramm
Mari Rich
Josh Robertson
Gregory K. Robinson
Albert Rolls
Margaret E. Roush
Liliana Segura
Olivia Jane Smith
Brian Solomon
Claire Stanford
Luke A. Stanton
Kate Stern
Maria A. Suarez
Hope Tarullo
Aaron Tassano
Cullen F. Thomas
Hallie Rose Waxman
Lara Weibgen
Selma Yampolsky

Contemporary Biographies in Chemistry

Barton, Jacqueline K.

American chemist

Born: May 7, 1952; New York, New York

"Lovely . . . with beautiful shape and symmetry." That is how chemist Jacqueline K. Barton described DNA (deoxyribonucleic acid) nearly twenty years ago during an interview with Stuart Gannes and Julianne Slovak for *Fortune* (October 13, 1986). A half-dozen years before, Barton had begun conducting experiments with DNA, the molecule that is basic to nearly all forms of life; orchestrates the activities of living cells; contains the genetic instructions for growth, development, and reproduction; and is so long that it might be considered "the biological equivalent of a colossal run-on sentence," as Gannes put it. Unlike most DNA researchers, Barton is neither a biologist nor a biochemist; rather, she is a specialist in bioinorganic chemistry, the study of inorganic compounds or ions (those that do not contain carbon-hydrogen bonds) in living organisms. In her research, she has treated the DNA molecule "as if it were an inorganic crystal, such as a piece of a rock or a semiconductor in a cell phone," thus introducing "a radically different way to think about DNA," as K. C. Cole wrote for the *Los Angeles Times* (December 29, 1997). By her early thirties, Barton had already made discoveries deemed so important that, in 1985, she became the first woman to win the National Science Foundation's Alan T. Waterman Award, the federal government's highest honor for scientists thirty-five years old or younger. Those discoveries were products of her pioneering creation and use of chemical "tools," made with metallic compounds, to recognize and modify specific sites on the DNA molecule. Then and in the years since, she has illuminated aspects of DNA's structure and some of the ways in which the molecule functions, including its conducting of electricity (which previously most scientists in her field had dismissed as impossible) and abilities to maintain and repair itself. According to a press release on the website of the California Institute of Technology (Caltech) announcing her election to the National Academy of Sciences (May 1, 2002), Barton's work "provides a completely new approach to the study of DNA structure and dynamics and may be critical to understanding the chemical consequences of radical damage to DNA within the cell." As the National Science Foundation noted on its website, her findings have "important implications for drug design and for the theory of gene expression." "Perhaps because I'm a woman and now getting some attention I'm a role model," Barton told Steven Waldman for the States News Service (May 15, 1985) after receiving the Waterman Award. "That sort of thing makes me a little uncomfortable." However, she added, "young women need to see that women are doing science and doing it well." Barton taught and worked at Columbia University in New York City from 1983 until 1989, when she joined the faculty of Caltech in Pasadena. There, she holds the title of Arthur and Marian Hanisch Memorial Professor of Chemistry. Barton has published as author or coauthor more than 250 scientific papers and book chapters. Her many honors include seven awards from the American Chemical Society, election to the American Academy of Arts and Sciences and the National Academy of Sciences, and a 1991 MacArthur Foundation Fellowship, popularly known as the "genius grant."

Early Life and Education

Jacqueline K. Barton was born on May 7, 1952, in New York City. She attended a private high school, the Riverdale Girls School (now part of the co-educational Riverdale Country School), in the Bronx, at a time when, as she told Natalie Angier for the *New York Times* (March 2, 2004), "young girls didn't take chemistry"; indeed, her school did not offer a course in that subject. She enrolled in a chemistry class for the first time as a student at Barnard College (which was associated with and later became a division of Columbia University) in New York City, which she entered in 1970. Barnard was then a women's school, and as an undergraduate Barton "learned that there wasn't anything strange about a woman going into science," as she told Steven Waldman. She majored in chemistry and graduated with a bachelor's degree summa cum laude in 1974.

Barton next began graduate studies in physical chemistry at Columbia; it was only then that she noticed the prevalence of men in her field. But increasing numbers of women were becoming scientists, and she encountered little discrimination; in her experience, she told Waldman, scientists' work "generally speaks for itself." She was a National Science Foundation predoctoral fellow from 1975 to 1978. Barton chose as her PhD adviser the bioinorganic chemist Stephen J. Lippard, after he aroused her curiosity about a substance in which he himself had grown interested—a compound of platinum called cis-diamminedichloroplatinum, also known as cis-platinum or cisplatin, which had proven to be highly effective in treating certain types of cancers. Barton struck Lippard as "clearly a very bright student, quite strong mathematically. Pure algebra, quantum mechanics—she was really good at that," as he told Will Hively for *Discover* (October 1994). Lippard suggested that she try to determine the structure of platinum blue, the compound formed by cis-platinum and uracil (the latter of which is very similar to thymine, one of the basic components of DNA)—"no easy task," as Hively wrote: "First she had to produce a crystal large enough to study; then she had to figure out the location of atoms in the crystal by analyzing the patterns produced by X-rays shot through the crystal." In what Lippard described to Hively as "a major breakthrough," Barton succeeded. The platinum-blue molecule, she determined, contained on its periphery several exposed platinum atoms that could potentially bond with other atoms—including those in components of DNA. Depending on the location, their bonding with the DNA molecule might hinder the further reproduction of cancer cells. Thus, as Hively wrote, "Barton's discovery opened up a whole new field of anticancer compounds." "Crystalline Platinum Blue: Its Molecular Structure, Chemical Reactivity, and Possible Relevance to the Mode of Action of Antitumor Platinum Drugs," by Barton and Lippard, was published in the *Annals of the New York Academy of Sciences* in 1978. Barton earned a PhD in inorganic chemistry in 1978. During the next year, supported by postdoctoral fellowships, she worked with the biophysicist and biochemist Robert G. Shulman at Bell Laboratories and then, in 1980, at Yale University in

New Haven, Connecticut. From 1980 to 1982 she taught chemistry at Hunter College, a division of the City University of New York. She joined the faculty of Columbia in 1983 and remained there for the next six years, during the last three with the title professor of chemistry.

Life's Work

When Gary Taubes, writing for *Science Watch* (January/February 1997), asked Barton how she would describe the "overriding theme" of her research, she responded, "My interest is in using chemistry to ask molecular questions about biological systems—to explore on a chemical level the relationship of structure to function." The goal of Barton's work since the 1980s has been to gain a better understanding of DNA, a nucleic acid whose structure was discovered in 1953 to be a double helix. In humans, DNA is present in twenty-three pairs of chromosomes (each parent having provided one set of twenty-three) located within the nucleus of almost every cell. (Among the few exceptions are mature red blood cells, which do not have nuclei, and reproductive cells, or gametes—sperm and eggs—each of which has only one set of twenty-three, not two.) Each chromosome contains one DNA molecule. Infinitesimally narrow but about six feet long, the molecule is extremely tightly coiled; if stretched to its full length, it would resemble a ladder that has been twisted along its midline millions of times. Each side of the ladder (also referred to as a strand or a backbone) looks like a necklace strung with millions of linked beads, with each bead composed of a phosphate molecule bonded to a sugar molecule (the sugar being deoxyribose). The rungs, or steps, of the ladder are all of equal length, and the outer ends of each rung are connected to one of the sugar-phosphate "beads" that lie along each strand. The ladder has two types of rungs, with each type made up of a pair of molecules called bases; there are four bases, each of which is a different arrangement of carbon, hydrogen, oxygen, and nitrogen atoms. One type of rung, the A-T pair, consists of the base adenine (A) weakly linked by two hydrogen bonds to the base thymine (T); the other type, the C-G pair, consists of the base cytosine (C) weakly linked by three hydrogen bonds to the base guanine (G). In normal DNA, A connects only to T, and C connects only to G. Thus, one side of the ladder is the complement of the other: if a sequence of nine bases on one side of the ladder is CCTAGTTGG, the corresponding sequence of nine bases on the other side of the ladder must be GGATCAACC. The chemical "direction" of one side of the ladder is the opposite of the other's; the two strands are said to be "antiparallel," and each plays a different role in the cell. Among the findings of the fed-

> **"Young women need to see that women are doing science and doing it well."**

eral government's ongoing Human Genome Project is that the number of base pairs (the "rungs") in a DNA molecule ranges from nearly 47 million in chromosome 21 to more than 245 million in chromosome 1. A base plus its connected sugar and phosphate molecules is called a nucleotide. A gene is a connected series of nucleotides; every human's approximately 20,000 to 30,000 genes (the precise number has not yet been determined) range in length from fewer than one thousand nucleotides to several million; the average number is about three thousand. Genes are described in terms of codons, each codon being a sequence of three of the four bases (TTT, TCC, CTA, CTG, etc.); the total number of codons is sixty-four (the four bases combined three at a time; $4^3 = 64$). Every codon contains information connected with the creation of one of twenty standard amino acids (aspargine, histidine, leucine, tryptophan, etc.), the components of proteins, from which muscle, nerves, blood vessels, bone, and everything else in the body is made. (Since there are only twenty, each amino acid is associated with more than one codon.) Each gene is located in a specific position on a particular chromosome. Only about 2 percent of the nucleotide sequences of human DNA are genes; the functions of the rest—known variously as junk DNA or noncoding DNA—are only beginning to be elucidated. Although the media and scientists, too, refer to "the" human genome, "a" human genome would be more accurate, because no two individuals have precisely the same DNA; according to the website Genome.gov, a comparison of the DNA of individuals will show variations of one nucleotide per thousand nucleotides, on average. (It is those variations that enable forensic scientists to identify criminals from blood or other body tissues.)

Practical Applications of Barton's Research

In 1979 scientists discovered that in some DNA molecules, dubbed Z-DNA, portions of the ladder "twist jaggedly to the left rather than spiraling smoothly to the right," as Julie Ann Miller wrote for *Science News* (April 20, 1985). At the same time, researchers were trying to recognize patterns in the nucleotide sequences and to demarcate the beginnings and ends of individual genes. Barton approached that problem from the standpoint of inorganic chemistry; as she told *Fortune*, "At some level one gene must be chemically distinguishable from another. Chemistry lets you look at a biological problem in a nice, simple way." To assess such chemical distinctions, she created compounds using atoms of transition metals, whose two outermost shells of electrons are similar and easily form chemical bonds; among the transition metals, she chose rhodium, ruthenium, and others among the so-called platinum metals. She then custom-made mirror-image pairs of compounds (called complexes), octahedrons that had an atom of the metal at their cores. She likened the shapes to a pair of mirror-image propellers; one propeller of each pair had a left-handed form, and the other a right-handed form. When she exposed both normal DNA (called B-DNA) and Z-DNA to the complexes, she discovered that only the right-handed propellers could slip between nucleotides in B-DNA, while the left-handed propellers behaved similarly with Z-DNA. Moreover, she could pinpoint the precise locations of such activity, because the metallic complexes—her "little probes," as she called them—emitted light when they found comfortable lodgings. "Take ruthenium," she explained, as quoted in *Fortune*, "When it luminesces, it really shines. It's like DayGlo. And it actually luminesces even more intensely when it is bound to DNA. When something happens, you know because the probe changes color." "That's part of the fascination," she added. By that method Barton saw that the complexes had bonded with the DNA molecules near the ends of genes—"sites where the Z-DNA is likely to exert gene control," as Julie Ann Miller wrote. Barton told *Fortune*, "We now have a little probe that is uniquely suited to detecting structural variations local to a site on the DNA strand." One potential benefit of such chemical tools lies in the treatment of disease. "Say you make an anti-HIV drug and you want to get it to bind to a certain spot on the DNA to stop cell division, for example," Barton told Cole. "So you take your drug, you attach a light switch. Then when it snuggles up in the DNA, when it's safe, it lights up. It's a good tracer. And it's nonradioactive."

In 1985 the National Science Foundation recognized the immense value of Barton's work by naming her the winner of that year's Waterman Award. Barton used the prize money ($100,000 each year for three years) to pay for new lab equipment and a larger staff. In the fall of 1989, she left Columbia and accepted a professorship at the California Institute of Technology. In 1991 she won a $250,000 MacArthur Foundation Fellowship. Two years later she was elected to the board of directors of the Dow Chemical Company, one of the world's leading manufacturers of chemicals. "For chemists and chemical engineers in industry and academia, it is important . . . that we find ways to enhance the public understanding of issues involving chemistry," she said, as quoted in *PR Newswire* (January 14, 1993).

Immeasurably aiding Barton in her work (and making possible the Human Genome Project) was the invention in the early 1980s by Leroy E. Hood (who was then at Caltech) of the DNA-sequencing machine, which can rapidly manufacture large quantities of synthetic DNA according to programmed instructions. About the size and shape of a microwave oven, the automated DNA synthesizer strings together fragments of genes to manufacture DNA whose base-pair sequences, unlike those in naturally occurring strands, are known precisely. With large quantities of such DNA at her disposal, Barton conducted research into the electrical conductivity of DNA and the ways in which the strand repaired itself. The latter was addressed in the abstract "Oxidative Thymine Dimer Repair in the DNA Helix," written by Barton and her colleagues Peter J. Dandliker and R. Erik Holmlin and published in the journal *Science* (March 7, 1997). Thymine dimers result from the linking of two thymine bases, an abnormality that can be caused by overexposure to ultraviolet light (a component of sunlight). Their linkage leads to mutations when DNA replicates itself, which in turn may lead to the uncontrolled cell growth that characterizes cancer. In the experiment described in *Science*, a synthetic rhodium compound of Barton's design was bonded to one or the other end of a DNA strand that contained a thymine dimer some distance from its ends. In the presence of normal visible light, the abstract explained, the compound "catalyzed" the "long-range repair," which was "mediated by the DNA helix": that is, the rhodium compound, though at some distance from the thymine dimers, stimulated the DNA strand to repair itself and thus return to its normal state. "What I think is exciting is that we can use the DNA to carry out chemistry at a distance," Barton said, according to *City News Service* (March 6, 1997). "What we're really doing is transferring information along the helix."

Overexposure to ultraviolet light or gamma rays causes the displacement or loss of electrons (a process known as oxidation) at bonding sites on DNA molecules. The displaced electrons, Barton discovered, could then travel to the spot on the DNA most susceptible to damage. (This was found to be the spot where multiple guanine nucleotides were adjacent to one another—"GG" and "GGG" sites.) Damaged GG sites, and the mutations they spawned, had already been believed to play an important role in cancer. As Jim Thomas explained in *New Scientist* (February 7, 1998), "If researchers can locate vulnerable sites and understand how harm is done, they can then look for ways to protect DNA from damage." The discovery of electron travel along the DNA strand is among the highlights of Barton's career. "I probably didn't realize what a crazy idea this was to some scientists," she told K. C. Cole. She also told Cole, "When you rattle the foundations [of science], people get nervous. But at a place like Caltech, it's okay to go out on a limb. It's okay to have an idea proved wrong. I'm supposed to be pushing the boundaries."

In 2001 Barton founded her own company, GeneOhm Sciences, to develop nucleic-acid-based tests "to detect and identify infectious agents and genetic variations" much more rapidly than is possible with other methods and thus enable physicians to begin treatment far sooner, according to the firm's website. Based in San Diego, California, GeneOhm acquired a Canadian company, Infectio Diagnostic, in 2004, and was itself purchased by BD (Becton, Dickinson and Co.) in early 2006. To date, GeneOhm has developed a nucleic-acid-based test for a type of streptococcus bacteria and another for detecting the presence of a virulent antibiotic-resistant bacterium that has often caused outbreaks of the infection in hospitals. The latter test is 96 percent accurate and takes two hours, rather than the two to four days necessary with other tests.

In one measure of Barton's influence, by the end of 1996, Barton's paper "Metals and DNA: Molecular Left-Handed Complements," in *Science* (Vol. 233, 1986), had been cited nearly 250 times in journal articles by other scientists. Barton was named chair of the Division of Chemistry and Chemical Engineering at Caltech in July 2009. In 2011, Barton received the National Medal of Science; the National Science Foundation cited her for the "discovery of a new property of the DNA helix, long-range electron transfer, and for showing that electron transfer depends upon stacking of the base pairs and DNA dynamics. Her experiments reveal a strategy for how DNA-repair proteins locate DNA lesions and demonstrate a biological role for DNA-mediated charge transfer." Barton currently serves as chair of the Environment, Health, Safety, and Technology Committee at Dow Chemical. She also continues to teach and lead a research group at Caltech.

Further Reading

Angier, Natalie. "Scientist at Work: Jacqueline Barton." *New York Times* 2 March 2004: F1. Print.

Barton, Jacqueline K. "Coping with technological protectionism." *Harvard Business Review* 6 (November/December 1984). Print.

Cole, K. C. "Putting DNA to Work as a Biomedical Tool." *Los Angeles Times* 29 Dec. 1997. Web. 12 Aug. 2012. <http://articles.latimes.com/1997>.

Hively, Will. "Married to the Molecule." *Discover* 1 October 1994. Kalmbach P. Print.

Jacqueline K. Barton. Chemical Heritage Foundation, Fall 2002. Web. 13 August 2012. http://www.chemheritage.org/discover/online-resources/.

Belcher, Angela

American chemical engineer

Born: June 21, 1967; Texas

Angela Belcher has won international renown as a scientist who thinks small. She began to focus on the infinitesimal as an undergraduate, when she realized that she "really liked molecules," as she said in an undated interview for the Texas Workforce Commission (TWC) website. Belcher is a world leader in nanotechnology, a science—named in 1974—in which the primary unit of measurement is the nanometer: one billionth of a meter, or about ten times the radius of a hydrogen atom or approximately 1/100,000th the width of a human hair. Every four or five years, Belcher has said, she likes to learn a "completely new field"; as a nanotechnologist, or, more specifically, a bionanotechnologist, she has drawn upon her knowledge of many disciplines, among them organic and inorganic chemistry, organic-inorganic interfaces, biochemistry, molecular biology, materials chemistry, biomaterials, biomolecular materials, genetic engineering, and electrical engineering. She and others in her field have noted that nature—specifically, biological nature—is world's most accomplished nanotechnologist, and in her research Belcher has studied "how nature makes materials," as she put it at a symposium called "Nanoscience—Where Physics, Chemistry, and Biology Collide," sponsored by the Australian Academy of Science, as quoted at Science.org.au (May 2, 2003). "We like to apply the ideas that nature has already used to make materials" to make other materials, ones "that we like to say nature hasn't had the opportunity to work with yet," as she said at that symposium. In another description of her approach, she told Elizabeth Corcoran for *Forbes* (July 23, 2001) that she and her research team are "working out the rules of biology" in a realm absent from the natural world. "The kinds of things that nature does that we like to do are basically self assembly, molecular scale recognition, nanoscale regularity, and self correction," she said at the nanoscience symposium. "One of our goals is to have biological molecules that can actually grow and assemble electronic materials . . . on the nanoscale and also act like enzymes and self correct. So as they start to grow the materials, if there is a mistake they would be able to correct themselves." Building on what she discovered as a doctoral candidate about the processes by which abalones use proteins and minerals to form their shells, Belcher invented techniques for coaxing viruses harmless to humans to create crystalline films and minuscule inorganic wires and battery anodes that are far smaller than those currently in use. Moreover, she created them without using the tremendous heat and toxic chemicals required by traditional manufacturing processes, such as photolithography, the means by which semiconductor devices are currently fabricated. Belcher's discoveries and inventions may lead to the creation of new generations of unprecedentedly small circuit boards and solar cells and lighter, more durable construction materials.

Belcher worked at the University of Texas at Austin from 1999 until 2002, when she joined the faculty of the Massachusetts Institute of Technology (MIT), in Cambridge. Since 2005 she has held the title of Germeshausen Professor of Materials Science and Engineering and Biological Engineering at MIT. Her many honors include a Presidential Early Career Award for Science and Engineering (2000), a Harvard University Wilson Prize in Chemistry (2001), a World Technology Award (2002), and a MacArthur Foundation Fellowship (2004), the last of which is popularly known as the "genius grant." In a conversation with Candace Stuart for *Small Times Magazine* (September 24, 2004), one of Belcher's mentors, Evelyn L. Hu, a professor of electrical and computer engineering at the University of California at Santa Barbara, said of her, "Angie has a special view of a way to change the world by thinking across disciplines and boundaries that is truly extraordinary. She has continuously looked into the future. She's very much in the excitement of the present but [has] always looked at the broader aspect."

Early Life and Education

A native of San Antonio, Texas, Angela M. Belcher was born on June 21, 1967. Readily available sources contain no information about her family and virtually none about her early years. In an undated interview for the TWC Web site, she said that as a twelve-year-old she planned to become a physician, and as a high-school student, she read medical books and accompanied doctors on their rounds at a hospital associated with Rice University, in Houston. She attended Santa Barbara City College (SBCC) in California from 1986 to 1988, supported in part by a William Olivarius Scholarship in Biology; while at the school she also won the SBCC President's Scholarship Award and the Outstanding Student Award in the Biological Sciences. She later transferred to the College of Creative Studies at the University of California at Santa Barbara (UCSB), which encouraged students to design their own curricula. "We could add or drop classes up to the week before finals, so that we could sign up for what we thought sounded interesting and after a few weeks pick what really was," she recalled in June 2003, in a commencement address presented at the College of Creative Studies and posted on the school's website. She concentrated in molecular biology and earned a bachelor degree in creative studies with highest honors from UCSB in 1991. During her undergraduate years she had served as an intern (1988) in gravitational and space biology at the Kennedy Space Center, a division of the National Aeronautics and Space Administration (NASA) in Florida; as a student researcher (1988–89) at the Center for the Study of Evolution and the Origin of Life and at the Plant Biochemistry Laboratory, both at the University of California at Los Angeles (UCLA), in the latter of which she studied metabolic pathways of a plant hormone called gibberellic acid; and as a student researcher (1989–91) at UCSB's Plant Molecular Biology Laboratory, where she investigated activity at the molecular level in the formation of alfalfa root nodules. During the summers of 1989 and 1990, as a UCSB field researcher, she studied the foraging behavior of wild honey bees on Santa Cruz Island. In pursuing a graduate degree in chemistry, also at UCSB, Belcher studied the processes involved in the formation of a natural biocomposite material—specifically, the shell of the abalone, a kind of mollusk. (Composites are materials whose constituent parts act together but retain their identities; the parts are physically distinguishable and do not dissolve or otherwise completely blend with one another. A biocomposite material contains organic as well as inorganic components and can grow without outside intervention. In addition to seashells, examples of biocomposites include bone, teeth, and ivory.) According to malacologists (experts on mollusks), at one time abalones had no shells; many millions of years ago, their internal proteins began to use calcium in seawater to produce shells composed of protein (2 percent) and calcium carbonate (98 percent). By means of chemical bonding, the proteins, working on the nanoscale at intervals of a millionth of a meter (called a micron), create hexagonal crystal-like tiles with carbonate particles; as more are created, the tiles become stacked, and the stacks form brick-like walls. A layer of protein "glue" binds each tile with the ones above and below it but does not bind the edges of horizontally adjoining tiles. The resulting walls are harder than synthetic ceramics and three thousand times stronger than calcium carbonate. "The hardness and luster [of the shell] is a function of the very, very uniform structure of calcium carbonate, deposited a molecule at a time" under the control of the proteins. "That's also what makes pearls," Belcher—who as a graduate student created pearls, albeit flat ones—told Spence Reiss for *Technology Review* (February 1, 2005). As in such other natural materials as spider silk and horses' hooves, the organic and inorganic molecules in abalone shells are arranged in a highly durable configuration. "If you look at how nature makes these materials, it makes them just fine," Belcher told Cindy Tumiel for the *San Antonio (Texas) Express-News* (June 8, 2000). "We want to be able to use that same exacting assembly on materials that nature hasn't evolved to have these kinds of interactions."

Life's Work

Belcher won a Parson's Fellowship in Materials Research (1994–95), a Distinguished Research Presentation Award at PacificChem in 1995—a conference organized by the American Chemical Society, and an Outstanding Chemistry Graduate Student Award from the American Institute of Chemists (1996). She earned a PhD from UCSB in 1997. For the next two years, she worked as a postdoctoral fellow there. In 1999 she received young-investigator awards from the US Army and Du Pont. That same year she joined the faculty of the Department of Chemistry and Biochemistry at the University of Texas at Austin. She was given her own laboratory and became the director of a group of about eighteen researchers.

In *Nature*, in an article entitled "Selection of Peptides with Semiconductor Binding Specificity for Directed Nanocrystal Assembly" (June 8, 2000), Belcher, Evelyn Hu, and three of their colleagues announced that they had successfully bonded viruses to five inorganic materials commonly used in semiconductors, including gallium arsenide, indium phosphide, and silicon. Viruses are so small that they can be seen only with electron microscopes, not ordinary microscopes (in which specimens are illuminated by light). Unlike a typical cell or a single-celled organism, such as a bacterium—which has a nucleus and various other small components surrounded by cytoplasm within an outer wall, and which can reproduce on its own—a virus consists only of genetic material (one or another of the two kinds of nucleic acid) enclosed in a protein wall; it can reproduce only when it is inside a suitable living cell. Belcher used viruses known as bacteriophages, which are harmless to mammals and reproduce rapidly in bacteria; each bacteriophage was about 6.6 nanometers wide and 880 nanometers long. Belcher bred in bacteria and then genetically manipulated roughly a billion individual viruses, each identical except for the protein molecules (called peptide amines) at one end. She next covered them with semiconductor particles—gallium arsenide was one of the first to be tried—and then sifted out the relatively few virus molecules whose proteins successfully bonded with the semiconductor molecules. She made millions of additional copies of those viruses (by providing them with bacterial hosts) and then reexposed that batch to uncontaminated semiconductor material "under more chemically stringent conditions," as she explained in her nanoscience lecture. "We go through this process about seven times," she said in the same talk. "We think of it as a Darwinian kind of process: we are looking for the ones that survive under the conditions that we are interested in. . . . In my group we can evolve interaction for materials in about one week." She thus isolated viruses whose peptide tips "recognized" molecules of a specific semiconductor but did not bond with nearly identical molecules of the same semiconductor (that is, isomers, which have the same atomic formulas but differ in the molecular arrangement of their atoms) or with different semiconductors. In the following months Belcher engineered proteins on viruses that bonded with more than twenty different inorganic materials. In 2000 Belcher received a Presidential Early Career Award for Science and Engineering (which came

> **"I have the career I have always dreamed of.... My advice is to follow your passion. Find that idea or topic that you're so excited about, that you wake up in the middle of the night and wish it was morning so that you can try it out."**

with a $500,000 research grant) from President Bill Clinton for her work. Soon afterward Clinton officially launched the National Nanotechnology Initiative.

Belcher's next goal, as she described it in her nanoscience talk, was "to grow and assemble a material." At the fall 2001 meeting of the Materials Research Society, and later in *Science* (May 3, 2002), she announced that she and her colleagues had accomplished that feat—a significant step toward the realization of marketable bioelectronics. Using a process similar to the one described in the 2000 *Nature* article, the scientists combined a genetically engineered bacteriophage with zinc sulfide. In *Science* they described the resulting product as a "highly ordered composite material": a liquid crystal film that was "ordered at the nanoscale and at the micrometer scale into [approximately] 72-micrometer domains, which were continuous over a centimeter length scale"—that is, big enough to be grasped with tweezers. The proteins at the tips of the viruses were weakly magnetic, enabling Belcher and her graduate student Seung-Wuk Lee to create different structural patterns in the film by means of a magnetic field. "In one such film, . . . the viruses lined up in successive rows, like rows of pencils all with zinc selenide erasers at one end," Robert F. Service wrote for *Science* (December 21, 2001). "Other films have viruses lining up in either a zigzag pattern or all facing the same direction but not organized into orderly rows." After noting that polymer-based liquid crystal mixtures without living ingredients had already been integrated into laptop computer screens, Service wrote, "Although the viral liquid crystals aren't likely to displace their polymer counterparts anytime soon, they point the way forward in one of the hottest areas of materials science." The film was also durable: after seven months the viruses were still intact, which suggested that the film could be used for storage purposes—to preserve genetically altered DNA, for example, or as an alternative to refrigerating vaccines shipped overseas. "We're looking at these materials to grow and arrange electronic, magnetic, and optical materials for devices, displays, and sensors," Belcher told Kimberly Patch for *Technology Research News* (May 15, 2002).

Viral-Based Batteries

In the fall of 2002, Belcher shifted her research to the NanoMechanical Technology Laboratory at MIT, where she joined the faculties of two departments as the John Chipman Associate Professor of Materials Science and Engineering and Bioengineering. Most of the postdoctoral students who had worked with her at the University of Texas accompanied her. The aim of her next undertaking was to tailor viruses to function as battery anodes—a first step toward the creation of small, malleable, and efficient viral-based batteries. With funding provided by the Army Research Office Institute of Collaborative Biotechnologies, Belcher and her team of researchers first altered the proteins of a bacteriophage so that the virus collected molecules of cobalt-oxide and gold when dipped in a solution; the virus then assembled the molecules into a single-layer nanowire that functioned as a battery anode. Belcher's team then demonstrated the virus's battery power, using a traditional lithium cathode to complete the circuit. Her research goals include creating a lithium cathode using similar means. Once the creation of an entire battery circuit based on viruses is complete, Belcher hopes to market the product commercially. "These batteries are like Saran wrap," she told R. Colin Johnson for *Electronic Engineering Times* (April 17, 2006). "They are thin, flexible and can be bent into any shape, making them good for lightweight conformable applications." She also told Johnson, "Potentially, when we grow a lithium layer on the other side of the polyelectrolyte for the other cathode, we could use this material to make batteries as thin as 100 [nanometers]." Belcher has predicted that such a battery will store two or three times more energy for its size than standard batteries and will lead to the advent of portable battery-powered devices far thinner and smaller than existing ones.

In 2002 Belcher and Evelyn L. Hu cofounded Semzyme Inc., which was renamed Cambrios Technologies in 2004. Funded by several venture-capitalist groups and currently based in Mountain View, California, the company aims to develop commercial uses for Belcher's innovative techniques. In 2004, according to *Ascribe Newswire* (September 28, 2004), after Belcher learned that she had won a $500,000 MacArthur Foundation grant, she told a reporter that the money would serve as "a catalyst for exploring new ideas in my lab and, equally important, let me contribute more to my community through science outreach to kids." In 2010, Belcher was awarded the Eni Award for Renewable and Non-Conventional Energy for "using nature as a guide for designing and building new materials for energy applications."

Personal Life

Belcher lives near Cambridge, Massachusetts, with several dogs. She is a vegetarian, and her dogs, Rebecca Skloot reported in *Popular Science* (November 2002), "eat only homemade organic foods." "Sometimes I can't sleep because I have so many ideas," Belcher told the TWC interviewer, to whom she revealed that, by choice, she was working at least eighty hours a week. In her 2003 commencement address at UCSB, Belcher said, "I have the career I have always dreamed of.... My advice is to follow your passion. Find that idea or topic that you're so excited about, that you wake up in the middle of the night and wish it was morning so that you can try it out. Or that when you email a colleague in the middle of the night to bounce an idea off of them, they email you right back, because they're also up, passionately pursuing an idea."

Further Reading

Belcher, Angela. "Biologically Templated Photocatalytic Nanostructures for Sustained Light-driven Water Oxidation." *Nature Nanotechnology* 5 (2012): 340–44, 2010. Print.

---. "Fabricating Genetically Engineered High-Power Lithium Ion Batteries Using Multiple Virus Genes." *Science* 32 (2009): 1051. Print.

---. "Virus-templated Self-assembled Single-walled Carbon Nanotubes for Highly Efficient Electron Collection in Photovoltaic Devices." *Nature Nanotechnology* 6 (2011): 377–84. Print.

Corcoran, Elizabeth. "The Next Small Thing." *Forbes* 23 July 2001. Print.

"Global Warming's Big Thinkers: Angela Belcher." *Time Specials.* Time Inc., n.d. Web. 13 August 2012. <http://www.time.com/time/specials/2007/innovators/>.

Bertozzi, Carolyn R.

American chemist

Born: October 10, 1966; Boston, Massachusetts

Carolyn R. Bertozzi is a pioneer in the field of cellular engineering, which involves not the design of cell phones, as one might suppose, but the transformation of cells in the human body into powerful disease fighters or diagnostic tools. "Our primary goal is to take control of the cell surface," Bertozzi has said, as quoted by *ScienceDaily* (January 6, 1998). "We have begun to understand the bio-organic chemistry of cells well enough to treat cells like complex machines—to really do cellular engineering." Bertozzi is a professor of chemistry and of molecular and cell biology at the University of California–Berkeley (UC–Berkeley); she also holds the titles of staff scientist at the Ernest Orlando Lawrence Berkeley National Laboratory and investigator at the Howard Hughes Medical Institute, a nonprofit organization for biomedical research at UC–Berkeley. Bertozzi's research focuses on how interactions that occur on the surfaces of cells contribute to health and disease. "We're trying to move beyond the level of how biological molecules interact with each other," Bertozzi told an interviewer for an article that appeared on a UC–Berkeley College of Chemistry website in 1995. "We're looking at cells and multicellular systems. How do small molecules control the behavior of a whole cell, and how do they dictate the way a cell behaves within tissues and organs?" Bertozzi, who holds five commercial patents, cofounded Thios Pharmaceuticals, a California-based drug-development company, in 2000. Her many honors include a 1999 MacArthur Foundation "genius" grant and, in 2003, election to the American Academy of Arts and Sciences.

Early Life and Education

The second of the three daughters of William Bertozzi and the former Norma Berringer, Carolyn R. Bertozzi was born on October 10, 1966, in Boston, Massachusetts. Her father was a professor of physics at the

Massachusetts Institute of Technology. Her parents' enthusiastic promotion of careers in the sciences struck a chord with their eldest daughter as well as the middle one: Andrea Bertozzi is a professor of mathematics and physics at the University of California–Los Angeles. (An extremely precocious youngster, Andrea Bertozzi was considered "the smarter one in the family," as Pervaiz Shallwani wrote for the *San Francisco Chronicle* (June 23, 1999). When Carolyn Bertozzi learned that she had won a MacArthur Fellowship, she recalled to Shallwani, her "first reaction" was that the foundation had "got the wrong Bertozzi.")

Inspired by her high-school biology teacher, Bertozzi told Kevin MacDermott for *Chemical & Engineering News* (January 21, 2002), she chose biology as her major when she entered Harvard University, in Cambridge, Massachusetts, in about 1984; in her sophomore year she switched to chemistry. As an undergraduate she worked with Joseph Grabowski, one of her chemistry professors, on the design and construction of a photoacoustic calorimeter, a device for studying the excited states of molecules by means of sound waves produced by absorbed light. The project rigorously tested her capabilities, giving her "the chance to rise to the occasion, and in the end, accomplish more than I thought I could at that stage," as she told Kevin MacDermott. "This was a confidence-building experience of unparalleled proportion in my life." Bertozzi graduated from Harvard with a bachelor's degree, summa cum laude, in 1988. She spent the following summer at Bell Laboratories, working with the chemist Christopher E. D. Chidsey on another challenging assignment: to determine properties of electroactive alkanethiol monolayers on gold surfaces. (Modifying metallic surfaces in this manner has many potential applications in industry.)

Bertozzi next entered the doctoral program in chemistry at UC–Berkeley; her primary research involved the synthesis of carbohydrate-based molecules. From 1988 to 1991 she was a US Office of Naval Research predoctoral fellow. During that period she attended a seminar led by Laura L. Kiessling, a highly accomplished University of Wisconsin professor of chemistry and biochemistry, whose honors include a MacArthur fellowship. "It was an event I will not forget," Bertozzi told MacDermott, "because she was the first female chemist

I had seen give a research seminar, believe it or not." In 1993 Bertozzi earned a PhD in chemistry. During the next two years, as an American Cancer Society postdoctoral fellow at the University of California at San Francisco, she studied cell biology and immunology, thus broadening her expertise from molecular chemistry to biological systems.

> "We have begun to understand the bio-organic chemistry of cells well enough to treat cells like complex machines—to really do cellular engineering."

"At the time, a lot of people considered that a radical move . . . ," Bertozzi told an interviewer for the UC–Berkeley College of Chemistry website. "Actually, it turns out that biologists and chemists are not as different as we think. We have a lot in common once you break down the lingo barriers." At UC–San Francisco she collaborated with Steven D. Rosen, a professor of anatomy, in investigating the role of oligosaccharides, a group of carbohydrates, in relieving inflammation.

Life's Work

In 1996 Bertozzi joined the faculty of UC–Berkeley as an assistant professor of chemistry; concurrently, she was hired as a faculty associate in the Materials Sciences Division of the Ernest Orlando Lawrence Berkeley National Laboratory, a multidisciplinary scientific facility run by the US Department of Energy under the auspices of the University of California. In 1999 Bertozzi received two promotions: she was named the head of the Chemical-Biology Department at Lawrence and an associate professor at UC–Berkeley.

"The trouble with this job is there are so many fascinating problems to study," Bertozzi told the College of Chemistry interviewer. "At a place like Berkeley, where you have the opportunity to work with some of the best students in the country, you feel like you could do everything." Bertozzi's research concerns the biological functions of carbohydrate molecules located on the surfaces of cells. For living organisms, carbohydrates (compounds, such as starches and sugars, that are made of carbon, hydrogen, and oxygen) are both sources of energy and components of cells. The surfaces of most mammalian cells are covered by a thick layer of oligosaccharides, each of which consists of a chain of several simple sugars, or monosaccharides. Cell-surface oligosaccharides serve as signaling molecules—that is, they send messages to other cells in order to coordinate their functions to benefit the whole organism. Oligosaccharides can serve as attachment points for other cells, disease-causing organisms, and various molecules; they thus play a critical role in protection from sickness and disease, cell-to-cell communication, and the regulation of immune responses. The structures of oligosaccharides on a cell's surface change over time and with the development and transformation of the cell. Such transformation takes place continuously in the bone marrow to produce blood cells; it also occurs at damaged tissue sites, in the skin, and during the progression of cancer. The change from a healthy to a cancerous cell, for example, is preceded by a change in the oligosaccharides. The goal of Bertozzi's work is to better understand cell-surface interactions, as a means toward developing treatments for diseases.

Bertozzi's research can be divided into three main areas. The first involves developing strategies for engineering chemical reactions on cell surfaces. This may help in such procedures as the delivery of genes to targeted locations and the tagging of tumor antigens (molecules found only on tumor cells) so as to make them visible in, for example, MRI scans, thus providing a way to discover the presence of tumors. The second main area of Bertozzi's research involves the study of carbohydrate sulfotransferases, a group of enzymes that function within a cell to modify oligosaccharides in response to events outside the cell. (The modification takes place inside the cell; the sulfated oligosaccharides are ultimately displayed on the cell surface.) Such research may be useful in developing therapies to relieve inflammation. A third avenue of research focuses on the development of new methods for the synthesis of homogeneous glycoproteins—compounds consisting of both carbohydrates and proteins, in which the carbohydrates and proteins are identical in all the glycoprotein molecules—and mimetics, molecules whose features imitate those of glycoproteins. This area includes the production of new glycoprotein structures within cell-based systems, and the substitution of problematic links within glycoproteins with bonds that are easier to synthesize. ("Oligosaccharides are hard to make chemically," Bertozzi told *Current Biography*, "but some of their bonds present a greater challenge than others. We replace the difficult-to-make bonds with unnatural linkages that are easier to make.")

One of Bertozzi's most lauded accomplishments is her discovery of a way to attach artificial markers to cell surfaces. (A marker makes cells easier to see in diagnostic tests.) She did so by first speculating that if a cell ingested a synthetic sugar with chemical properties that differed from those of sugars already on the surface, the sugar might be incorporated into the cell-surface oligosaccharide. (Some molecules can move from within a cell to the surface of the cell.) The cell surface would thus be "marked" with the chemical properties of the synthetic sugar. Marking cells in this way provides scientists with an innovative way of targeting cells for cancer therapy, adhering cells to materials used in biomedical implants and reactors, and executing other operations. Such artificial markers may someday lead to the use of devices that warn of chemical or biological toxins in cellular environments.

In addition to the MacArthur Fellowship, Bertozzi has won nearly two dozen research or teaching awards, among them the Pew Scholars Award in the Biomedical Sciences (1996); the Prytanean Faculty Award (1998), from the Prytanean Women's Honor Society; the Beckman Young Investigator Award (1998); the American Chemical Society Award in Pure Chemistry (2001); and the Tetrahedron Young Investigator Award for Bioorganic and Medicinal Chemistry (2011). She has contributed to more than one hundred scholarly publications and, with Peng George Wang, coedited a textbook, *Glycochemistry: Principles, Synthesis, and Applications* (2001). She holds five commercial patents, one of which is for a material used in the manufacture of contact lenses, and sits on the advisory boards of several biotech and pharmaceutical companies, including Zyomyx, which makes tools for analyzing proteins. She is also the cofounder of the Emeryville, California, drug-development company Thios Pharmaceuticals. In 2010, she became the first woman to receive the prestigious Lemelson-MIT Prize. In 2011, Bertozzi became an elected member of the Institute of Medicine, one of the highest honors in the fields of health and medicine.

When Kevin MacDermott asked her to name her most significant achievement, Bertozzi answered, "I was most proud to have graduated my first PhD student and see her develop her own proposal as a postdoctoral fellow."

Further Reading

"Carolyn R. Bertozzi." *Bertozzi Research Group: University of California, Berkeley,* n.d. Web. 13 August 2012. http://www.cchem.berkeley.edu/crbgrp/.

Wang, Linda. "Carolyn Bertozzi Named Kavli Lecturer." *Chemical & Engineering News* 89.31 (1 August 2011): 48. Print.

Preussl, Paul."Cellular Engineers Design Custom cell Surfaces Able to adhere to Synthetic materials." *Berkeley Lab: Research News* 5 (January 1998). Print.

Shallwani, Pervaiz. "Bay Area Surprise Winners: UC Berkely Hits the Jackpot with Nation's Best crop of MacArthur Fellowship Recipients." *San Francisco Chronicle* 23 June 1999. Web. 13 August 2012. <http://www.sfgate.com/education/article/>.

Catlin, Don H.

American pharmacologist

Born: June 4, 1938; New Haven, Connecticut

In recent years, the use of steroids and other banned strength-and-endurance aids among athletes has become common, so much so that outstanding feats in sports often trigger suspicion. Don H. Catlin, a physician, is one of the founding fathers of drug testing in sports. Over the last quarter-century, he has developed numerous drug-identification techniques and helped lead the fight against athletes' use of performance-enhancing substances. Catlin is the founder (in 1982) and former director of the UCLA Olympic Analytical Laboratory, the first drug-testing lab in the United States to be accredited by the International Olympic Committee (IOC). Under his twenty-five-year stewardship, the lab grew to become the largest testing facility for performance-enhancing drugs in the world, processing approximately 40,000 urine samples a year and providing drug education and tests for many national and international sports organizations, including the US Olympic Committee, the National Collegiate Athletic Association (NCAA), the National Football League (NFL), and Minor League Baseball. During that time Catlin developed the groundbreaking carbon-isotope ratio (CIR) test, which determines whether testosterone in athletes' bodies is natural or has come from a prohibited performance-enhancing drug. He also led efforts to get the IOC to ban androstenedione and other male-hormone supplements. Catlin's greatest achievement came in 2003, when he cracked the code for the previously undetectable "designer" steroid tetrahydrogestrinone (THG), popularly known as "the Clear." That breakthrough led to the decoding of other steroids and played a key role in the Bay Area Laboratory Cooperative (BALCO) investigation, which linked such high-profile athletes as Barry Bonds, Jason Giambi, Marion Jones, and a slew of other professional athletes to use of a wide array of performance-enhancing drugs; Catlin was honored by the *Chicago Tribune* as sportsman of the year for his contribution to that investigation.

In 2005 Catlin became the founder and chief executive officer of Anti-Doping Research Inc., a nonprofit organization based in Los Angeles, California. There, he has continued to develop new testing methods, including one aimed at detecting the human-growth hormone (hGH). Commenting on the proliferation of drug use in sports, Catlin explained to Dan Shaughnessy for the *Boston Globe* (July 16, 2006), "People think of this as cops and robbers. That's not really it. I look at any positive test as a failure of the system. A failure to explain things. A lot of [those who test positive] are innocent. They get stuff that's contaminated and they get whacked even though they have no intention to cheat. Those guys don't bother me. We want to catch the cheaters. If you believe that sport is good for life and society and is worth preserving, you have to do something about it." He also said, "I think a clever crook can beat us. Once we get [the test for] growth hormone, they'll just shift a few degrees and start doing something else. There is so much money involved, there is so much desire to win, whether it be a pennant, a World Cup, the Tour de France. It doesn't matter. People want to win and they're going to do what they have to do and they might consider cheating. It's part of a way of life for an athlete. And that's tough." Catlin said to Jill Lieber Steeg for *USA Today* (February 28, 2007), "I can't think of anything more exciting than the Olympic model, where 220 countries in the world participate and every four years they send their best to compete against the best from other countries and the best man or woman wins. That's gorgeous. What could be nicer?" He told Steeg, "You should care about preserving something natural and beautiful."

Early Life and Education

Don H. Catlin was born on June 4, 1938, in New Haven, Connecticut, and was raised near the Berkshire Mountains of western Massachusetts. His father, Kenneth, was an insurance executive; his mother, Hilda, was a homemaker. An avid sports fan from youth, Catlin grew up "rooting for Ted Williams and the Boston Red Sox," as Steeg noted. One of his family's close friends was the pioneering surgeon and Yale University professor Gustaf Lindskog. "Private chats with him convinced me to go to medical school," Catlin told *Current Biography*. "That made sense to me because I feared boredom and medicine seemed to guarantee a lifetime of challenges." After graduating from high school, he attended Yale, in New Haven, where he received a bachelor's degree in statistics and psychology in 1960. Catlin then attended medical school at the University of Rochester in New York. He earned a doctoral degree in medicine in 1965.

Upon completing his medical studies, Catlin joined the US Army and became a specialist in internal medicine at the Walter Reed Army Medical Center, in Washington, DC. During that time he read an account of a Washington, DC., man who took drug addicts from the streets to his treatment center and helped them overcome their habits. "He had the right approach," Catlin recalled to Steeg. "He would dry them out, physically keep them away from drugs. He would literally hold them, then sing, play music, and talk to them." After the government threatened to close the center because it had no doctor on staff, Catlin offered to fill that position. On the strength of his reputation for working with addicts, he was put in charge of an army treatment program. He clashed with generals at the Pentagon over plans to jail soldiers who had turned to heroin to deal with the stress of the war in Vietnam. He recalled to Steeg, "I said, 'Wait, let's try the medical model.' But they locked 'em up anyway."

Life's Work

In 1972 Catlin joined the faculty of the University of California at Los Angeles (UCLA), beginning as an assistant professor in the Department of Pharmacology. He had been teaching at UCLA for nine years when the IOC recruited him to set up and run its drug-testing lab for the 1984 Summer Olympics, which were being held in Los Angeles. At the time only three labs in the world (in London, England; Paris, France; and Cologne, Germany) were conducting sports-doping research. While the IOC had started using limited doping tests at the 1968 Summer Olympics in Mexico City, not much other drug testing had been done. Catlin confessed to T. J. Quinn for the *New York Daily News* (June 11, 2006) that at first "the idea made no sense to me. Why would a young, healthy athlete want to take a drug? It seemed so stupid." He nonetheless took the assignment, in part because UCLA would keep the lab equipment after the Games. In 1982 he founded the UCLA Olympic Analytical Laboratory, the first anti-doping lab in the United States. At the behest of the IOC, he began educating athletes a year prior to the 1984 Olympics to give them a chance to experience the drug-testing process. Shortly afterward he tested a series of urine samples that, he discovered, contained a banned drug in decreasing amounts and realized that someone was using the program to figure out how long it took for the drug to disappear from athletes' systems. Catlin recalled to Steeg, "[The US Olympic Committee] wanted to do testing, and for us to tell them the answers, meaning whether their guys and girls test positive or not. But they didn't want any consequences. It would just be an educational experience. And already we could see that, well, what kind of education is this? We shouldn't be doing this. . . . That program stopped real fast."

After the 1984 Olympics Catlin approached Manfred Donike, a pioneer in drug testing for sports, about the lack of a test for Stanozolol, a synthetic anabolic steroid that Catlin had suspected athletes of using in the Los Angeles Games. Catlin said to Steeg, "[Donike] hands me a pill—Stanozolol—and says, 'Here, take this.' So, I took it, and I gave him my urine, and he couldn't find it." A year later Donike discovered a way to detect the drug. Throughout the 1980s Catlin made repeated pleas to the IOC to establish a code of ethics for Olympic drug-testing labs; the code was eventually adopted as a requirement for accreditation by the IOC. In 1988 Catlin persuaded the IOC to remove norethisterone, a substance common in birth-control pills, from its list of banned drugs; it had been barred because of its potential to generate a by-product also created by the banned anabolic steroid nandralone, which made it impossible for drug testers to know whether a female athlete had taken nandralone, a birth-control pill, or both. Catlin and his team of scientists found a method of determining which substance a female athlete had consumed. Also that year, in a development that rocked the sporting world, the Canadian sprinter Ben Johnson tested positive for Stanozolol, just days after winning the gold medal and breaking the world record in the 100 meters at the 1988 Summer

Olympics in Seoul, South Korea. Catlin, who was in Seoul when the scandal erupted, told Dan Shaughnessy, "That changed things a lot. It woke up a lot of sports people. When I came home from Seoul, my friends said, 'We're going to send our kids out for baseball. They don't have drugs in baseball.' That was 1988. People like sports and they don't like to believe how dirty it can be. So they put blinders on. It's hard for me to do that. When you get into my field, everything looks very different."

> **"We want to catch the cheaters. If you believe that sport is good for life and society and is worth preserving, you have to do something about it."**

In the 1990s Catlin developed the CIR test, which differentiates between testosterone produced naturally and that introduced through drugs. During that period, in efforts to set stricter drug-testing policies, Catlin suggested to the IOC that approval by the International Organization for Standardization be a prerequisite to IOC lab accreditation, an idea that the IOC adopted years later. Meanwhile, in 1998 Catlin got the IOC to ban other performance-enhancing drugs, including androstenedione and other male-hormone supplements. (Androstenedione, commonly known as "andro," was widely used throughout the 1990s by Major League Baseball players. Prior to its banning it could be purchased over the counter.)

The World Anti-Doping Agency and the US Anti-Doping Agency

The year 2000 saw the establishment of both the World Anti-Doping Agency (WADA) and the US Anti-Doping Agency (USADA). The former was created in the wake of the 1998 doping scandal involving Tour de France riders from the French Festina cycling team, which exposed many of the failings of the old drug-testing system. (The team was found to have doped their riders with everything from anabolic steroids to the red-blood-cell-boosting endurance drug erythropoietin, or EPO.) Up until then each of the various sports organizations had run its own anti-doping operation, which made it easier for athletes to cheat the system. The IOC created the WADA as an independent body that would establish a universal list of prohibited substances. The WADA does no testing of its own but comprises a global chain of accredited laboratories that do so. The UCLA Olympic Analytical Laboratory is currently one of thirty-five laboratories that conduct tests for the WADA. The USADA, by contrast, sends its own officers to collect urine samples and request tests from labs. Athletes who test positive for performance-enhancing drugs are then charged and punished by the agency. Catlin, for his part, has maintained that what he calls the cop-and-robber approach to drug testing is ultimately futile. "People are following this old model—run 'em down, chase 'em, find 'em, assume they are guilty, drag them into testing," he explained to Brian Alexander for *Outside* magazine (July 2005). "And athletes still get away with stuff, and I maintain you can get away with stuff with everybody looking *right* at you." He added, "The system has failed to deal with the problem. And it will fail now."

Despite his skepticism, Catlin continued to lead the charge against drug misuse in sports and developed a number of new tests. In 2000 the scientists in his lab learned to distinguish between natural testosterone and that produced by a drug made from yams. Then, prior to the 2002 Winter Olympics in Salt Lake City, Utah, Catlin used the French EPO test (developed by the scientist Françoise Lasne) to detect darbepoetin, a variety of EPO that increases endurance. That test helped expose three cross-country skiers who used darbepoetin while competing in the Games, including the gold medalists Johann Muehlegg of Spain and Larissa Lazutina of Russia. Later that year, after analyzing data from an athlete's urine sample, Catlin cracked the code for norbolethone, the first reported "designer" steroid, or synthetic steroid created to enable its users to evade existing drug laws. (The drug, aimed at increasing weight, was first synthesized by the pharmaceutical company Wyeth in 1966 but was never marketed, out of fear of harmful side effects.) Catlin explained to Steeg that his lab's discovery "proved beyond a shadow of a doubt that our suspicions were true, that there were designer steroids out there that we couldn't find and that people were getting a hold of them." In the same year, the US track cyclist Tammy Thomas was found to have used norbolethone and was banned for life from competing in the sport.

In 2003 Catlin cracked the code for another designer steroid. The circumstances of the discovery brought to mind "a B-movie mystery plot," as Brian Alexander noted. In June of that year, a syringe was mailed anonymously to the USADA. (The source was later revealed to be Trevor Graham, a former track-and-field coach who received a lifetime ban from his sport in 2008 for supplying many athletes with performance-enhancing drugs.) The USADA sent some of the syringe's contents to Catlin, who with his team of about forty researchers concluded that the substance was a previously undetectable, custom-made steroid, which they called tetrahydrogestrinone, or THG. Catlin then developed a special test for the drug. He noted to Steeg, "It was a wonderful time for us because it was a kind of project that took all the skills that are represented in this lab, in terms of PhDs, chemists . . . it took everybody."

The BALCO Investigation

Through his research Catlin was able to advise Jeff Novitzky, an agent of the Internal Revenue Service (IRS), who opened the BALCO case, leading to perhaps the biggest doping investigation in the history of sports. Headquartered in Burlingame, California, BALCO provided blood and urine analysis and food supplements for professional athletes. An investigation revealed that BALCO's owner and founder, Victor Conte, was working with a network of rogue chemists and steroid dealers to supply professional athletes with performance-enhancing drugs. Catlin, who had long suspected that bodybuilders and other athletes were being supplied with undetectable steroids from clandestine labs around the country, helped expose BALCO by testing more than 550 existing urine samples from athletes, twenty of which contained traces of THG. He was the first witness to testify before the grand jury in the BALCO case. Among the athletes named in connection with the BALCO case were the baseball players Barry Bonds, Jason Giambi, Gary Sheffield, Benito Santiago, and Jeremy Giambi; the football players Bill Romanowski, Tyrone Wheatley,

Barrett Robbins, Chris Cooper, and Dana Stubblefield; the boxer Shane Mosley; and the sprinters Dwain Chambers, Marion Jones, Tim Montgomery, Raymond J. Smith, and Kelli White; and many other track-and-field stars, including the shot putters Kevin Toth and C. J. Hunter. The scandal foreshadowed the findings of the Mitchell Report, the result of former US senator George J. Mitchell's twenty-one-month investigation into the use of anabolic steroids and human-growth hormone in Major League Baseball (MLB). The report, released on December 13, 2007, named eighty-five MLB players who were alleged to have used steroids or other banned drugs, including such superstars as Roger Clemens, Andy Pettitte, Miguel Tejada, and Eric Gagne. A number of other athletes have since come forth about their steroid use, most notably the baseball players Alex Rodriguez and Mark McGwire.

In 2004 Catlin cracked the code for madol (commonly known as DMT), the third reported designer anabolic steroid; in January 2010 the drug became a controlled substance, meaning that US federal law governs its manufacture, distribution, and use. In 2005 Catlin became the founder, president, and CEO of Anti-Doping Research Inc., a nonprofit research organization focused on performance-enhancing drugs. The organization has led efforts to uncover new drugs being used illegally by athletes and worked toward developing new tests to detect them. In 2007 Catlin stepped down from his directorship at the UCLA Olympic Analytical Laboratory to work exclusively with Anti-Doping Research. Over the last several years, the organization has attempted to produce an effective urine test for hGH, a hormone that has been used in the past primarily to treat abnormally short children. Many believe that the use of human-growth hormone is still common among professional athletes, making a successful hGH test highly sought-after by sports leagues worldwide. In the meantime Catlin and his researchers have been successful in developing tests for other drugs. Most recently, in late 2009, he and his team developed an equine test for the potent blood-boosting drug CERA (sold under the brand name Mircera), a long-acting form of EPO.

Catlin's organization has also worked to set up programs that will dissuade athletes at all levels from using performance-enhancing drugs. Catlin has long advocated the establishment of a volunteer program, through which athletes would submit to tests to create a set of "biomarkers" for each participant, showing what is normal and abnormal (regarding levels of testosterone or insulin, for example). The participants would have ongoing checkups to ensure that no banned drugs are in their systems and would be recognized as honest athletes. While the idea has been called far-fetched because it would rely on voluntary actions, Catlin believes that social pressure would ultimately lead most athletes to volunteer. He told T. J. Quinn, "We're fundamentally optimists; we have to be in this field. We're not proposing that there's a solution, we're proposing that it needs to be tried."

Brian Alexander described Catlin as tall and balding with "a handsomely craggy face." Catlin is professor emeritus of molecular and medical pharmacology at the UCLA David Geffen School of Medicine. He serves as chairman of the Equine Drug Research Institute's Scientific Advisory Committee and is a member of the Federation Equestre Internationale Commission on Equine Anti-Doping and Medication. In addition, he is a member of the International Olympic Committee Medical Commission. Catlin's wife, Bernadette, a nurse whom he met at UCLA, died of melanoma in 1989. He has two sons from that marriage: Bryce, a software engineer in San Francisco, and Oliver, who is vice president of Anti-Doping Research. According to Steeg, Catlin has received hate mail from athletes caught using drugs, and on one occasion his car was firebombed. Catlin has remained undeterred, however. Oliver Catlin told Steeg that since he started working with his father, he has become "fully aware of how dedicated he is to his field. He doesn't stop thinking about this stuff. He eats, drinks, and sleeps this stuff. It's his life. It's his cause. It makes him tick."

Further Reading

Bryant, Christa Case. "Gatekeeper for Clean Sports." *Christian Science Monitor* 5 (August 2008): 25. Print.

Catlin, Don. Interview by Peter Aldhous. "Confronting Drugs in Sports." *New Scientist* 8 August 2007: 2616. Print.

"Dr. Don Catlin Responds to SI Article on Lance Armstrong." *Outside: Adventure*, 27 January 2011. Web. August 13 2012. < http://www.outsideonline.com/blog/outdoor-adventure>.

Hall, Carl T. "How Sleuths Tracked a Mysterious Steriod." *San Francisco Chronicle* 26 October 2003. Print.

"Milestones in Don Catlin's Career." *USA Today: Sports,* 28 February 2007: C9. Print.

O'Keefe, Michael. "Scientist Don Catlin Understands NFLPA's Hesitation to Allow WADA to Test Players for HGH." *NY Daily News* 29 August 2011: 62. Print.

Shaughnessy, Dan. "Lab is helping baseball to clean up its act." *The Boston Globe* 16 July 2006: C p1. Print.

"UCLA Drug Lab Faces Loss of IOC Accreditation." *Los Angeles Times* 4 December 1997, n.d. Print.

Chauvin, Yves

French chemist

Born: October 10, 1930; Menin, France

On October 5, 2005, the Royal Swedish Academy announced that it had presented that year's Nobel Prize in Chemistry to Yves Chauvin, who shared the award with Richard R. Schrock and Robert H. Grubbs, "for the development of the metathesis method in organic synthesis," noting that the three winners' pioneering work had provided researchers with a powerful new means of synthesizing molecules. "The beauty of this is that it's a reaction that you can now guide very precisely . . . exactly as you like," Astrid Graeslund, the Nobel Committee secretary, told Rob Stein for the *Washington Post* (October 6, 2005). "If you were going to do this kind of reaction in the old way, it would involve many, many steps. . . . In every reaction, you make waste products that you have to discard," she said. "With this you can make a shortcut. You can make a reaction that causes very little waste and make only the product you want."

After learning that he had won the prize, Chauvin admitted to reporters that he was "embarrassed" by the deluge of media attention. "I knew that my research was important. I opened the way, but it is my American colleagues who also worked on my research who are enabling me to get this prize today," he remarked, as quoted by Stein. "It took thirty years of laboratory work to show that what I found was interesting."

Early Life and Education

One of five children, Yves Chauvin was born in Menin (spelled Menen in Flemish), a town in western Flanders, Belgium, on the French-Belgian border, on October 10, 1930. His parents were French, and his father, who served in both World War I and World War II, often worked long hours as an electrical engineer. Chauvin attended preschool in Flanders and later crossed the border each day to attend primary school in France. He completed his secondary and higher education in "various towns," according to an autobiography he wrote for the Nobel Prize's official website. "To be perfectly truthful, I was not a very brilliant student, even at chemistry school," he wrote. "I chose chemistry rather by chance, because I firmly believed (and still do) that you can become passionately involved in your work whatever it is." He earned his degree from the Lyon School of Chemistry, Physics, and Electronics in 1954. Due to a number of factors, among them his eventual military service, he never earned a PhD.

Life's Work

After graduating Chauvin took an industrial job that left him with little room to explore new ideas. "I discovered that this was a very common attitude among managers. They are afraid of anything new: 'Do what everyone else does and change as little as possible: at least we know it will work,'" he wrote. "It is the opposite of my way of thinking, which, I must admit, is a bit of an obsession! I have often got into arguments about it. My motto is, 'If you want to find something new, look for something new.'" In 1960, the same year he got married, he joined the Institut Français du Pétrole, in Rueil-Malmaison, France. He spent the majority of his career as director of research at the school. The research for which he was awarded the 2005 Nobel Prize in Chemistry focused on olefin metathesis, a chemical process in which molecules exchange groups of atoms.

Olefin metathesis is a form of organic synthesis, the method by which scientists create organic molecules. Organic synthesis forms the heart of organic chemistry, which deals with carbon-based molecules, the basic building blocks of all life on earth. Carbon-based molecules also comprise some nonorganic materials, such as plastics. Carbon atoms, which have four valence electrons, are able to form single, double, and triple bonds with atoms of such other elements as hydrogen and oxygen by sharing one, two, or three electrons, respectively. Carbon atoms can also bond with other carbon atoms, allowing them to form long chains, branched structures, and rings. Carbon-chain molecules that contain carbon atoms with double bonds are known as olefins (or alkenes). In olefin metathesis, as explained in a Royal Swedish

Academy of Sciences press release (October 5, 2005), which is posted on the Nobel Foundation website, "double bonds are broken and made between carbon atoms in ways that cause atom groups to change places. This happens with the assistance of special catalyst molecules."

Understanding Metathesis Reactions

Chemists have known about the existence of metathesis since the 1950s, when it was discovered by industrial researchers, but no one understood how the process worked. In particular, the special catalyst molecules that provoked the reaction remained a mystery. Chemists

> "My motto is, 'If you want to find something new, look for something new.'"

realized that unlocking that secret would be a tremendous benefit to science and industry. In the 1960s researchers discovered that olefin molecules played an essential role in the process, and hence termed it olefin metathesis. The nature of the reaction, however, remained unclear.

In 1971 (some sources state 1970) Chauvin and a student, Jean-Louis Herrison, published a report in which they outlined the "recipe" for metathesis reactions and demonstrated which compounds acted as catalysts in such reactions. Chauvin proposed that the catalyst was a metal carbene, a compound in which metal and carbon are held together by a double bond. He also produced experimental results that couldn't be explained by any previous theory of the mechanism. According to the Royal Swedish Academy in the supplementary information to its October 2005 press release, as posted on the Nobel website, Chauvin's mechanism can be viewed as a dance "in which the 'catalyst pair' and the 'alkene pair' dance round and change partners with one another. The metal and its partner hold hands with both hands and when they meet the 'alkene pair' (a dancing pair consisting of two alkylides) the two pairs unite in a ring dance. After a while they let go of each other's hands, leave their old partners, and dance on with their new ones. The new 'catalyst pair' is now ready to catch another dancing 'alkene pair' for a new ring dance or, in other words, to continue acting as a catalyst in metathesis."

"Chauvin's mechanism explained at one stroke all earlier outcomes of olefin metathesis," the Swedish Academy wrote. K. C. Nicolaou, a chemistry professor at Scripps Research Institute and the University of California, San Diego, told A. Maureen Rouhi for *Chemical & Engineering News* (December 23, 2002) that Chauvin's work gave the field "a chance to move away from its state of alchemy, although it took several years before the mechanism was experimentally supported and widely accepted." Chauvin recalled to Rouhi that his research had been inspired by three papers published in 1964—one of which was by Ernst Otto Fischer, a scientist at the Technical University of Munich, in Germany, who won the Nobel Prize in Chemistry in 1973. "Apparently, these papers had nothing in common," Chauvin explained to Rouhi. "But for me, they were a revelation." He added, "With the paper of Fischer, I felt that these species could be metal carbenes."

Although Chauvin's work did not produce immediate consensus on the role of metal carbenes, it was soon supported by the experimental results of other researchers. In a statement praising Chauvin, posted on the CPE Lyon website, J. M. Basset, the director of the Surface Organometallic Chemistry Laboratory of CPE Lyon and the Centre National de la Recherche Scientifique (CNRS), noted: "In the 1970s [Chauvin] proposed metallocarbene intermediates [as catalysts for the metathesis process], and it was only several years later that such species were isolated. His wide range of activity covered numerous aspects of homogeneous catalysis and polymerisation: olefin and diolefin oligomerisation, carbonylation, synthesis of natural and synthetic alpha amino acids by asymmetric catalysis, and the chemistry of the rare earth elements."

The next advancement in metathesis—developing the actual catalysts to be used in the process—was made by Richard R. Schrock, an American chemist at the Massachusetts Institute of Technology, in Cambridge, Massachusetts. In 1990 Schrock developed a successful catalyst using the metal molybdenum; it has since come to be known in the field as a Schrock carbene. Two years later Robert H. Grubbs, a scientist at the California Institute of Technology, in Pasadena,

California, used the metal ruthenium to develop an even more well-defined catalyst, one that was less reactive to oxygen and moisture, which meant it could be used in the open air.

Olefin metathesis has since been found to have uses in a number of fields, particularly in the development of new pharmaceuticals and advanced plastics. According to Paul Rincon for BBC News (October 5, 2005), "Benefits to arise from [metathesis] include advanced herbicides, additives for polymers and fuels, and research into new treatments for bacterial infection, cancer, Alzheimer's disease, arthritis, migraine and HIV." Commentators have also hailed the development of metathesis reactions as a leap forward in "green" chemistry, which advocates environmentally friendly methods of producing new chemicals. The process minimizes the production of hazardous byproducts. It is also simpler to use and more efficient than previous production methods of organic synthesis, requiring fewer resources and reducing the number of steps involved.

Chauvin serves as the emeritus director of research at the Surface Organometallic Chemistry Laboratory of CPE Lyon/CNRS. He has continued to contribute to the development of homogeneous catalysts for the metathesis reaction process. Chauvin is also credited as the originator of several important petrochemical processes, including the Dimersol process for adding propene to oil and the Alpha Butol process that converts ethane to butane-1. He has recently pursued developments in homogeneous catalysis in molten salts, a field of research that allows very selective homogeneous catalytic reactions to occur while granting an easy separation of the transition metal from the reaction mixture.

Chauvin lives in Tours, France, and is recently widowed. He is a member of the French Academy of Sciences, from which he received the Clavel-Lespiau Prize in 1990. He is also the recipient of the Carl Engler Medal from the German Scientific Society for Coal and Petroleum Research (1994).

Further Reading

Chang, Kenneth. "3 Win Nobel on Changing Pieces of Molecules." *New York Times* 6 October 2005: A35. Print.

Rincon, Paul. "Three Share Chemistry Novel Prize." *BBC News*, 5 October 2005. Web. 13 August 2012. Print. Web.

Rouhi, Maureen A. "Olefin Metathesis: The Early Days." *Chemical & Engineering News*. 23 (December 2002): 34–38. Print.

Stein, Rob. "American, French Chemist Win Novel." *Washington Post* 6 October 2005: A12. Print.

Chizmadzhev, Yuri Aleksandrovich

Russian physical chemist

Born: December 15, 1931; Russia

For almost forty-five years, Yuri Aleksandrovich Chizmadzhev has worked in the field of electrochemistry, the scientific discipline that deals with chemical changes produced by electrical action and the production of electrical action by chemical changes. He has been a protégé of many seminal figures in the field, including Alexander Naumovich Frumkin, Veniamin Grigorievich Levich, and Revaz Romanovich Dogonadze. His work, particularly his study of cell membranes and skin pores and the diffusion of chemicals through those pores, is of commercial and medical importance, because of the viability of delivering needed drugs through the skin using various methods.

Early Life and Education

Yuri Aleksandrovich Chizmadzhev was born on December 15, 1931. He attended the Moscow Institute of Engineering Physics, graduating in 1956. He then joined the electrochemistry department at the Institute of Physical Chemistry, which was later renamed the Institute of Electrochemistry, and is now known as the A. N. Frumkin Institute of Electrochemistry. The Frumkin Institute is part of the Russian Academy of Sciences, an elite research and educational institution in Moscow that had been founded by Peter the Great in 1724. (It was originally called the Imperial Academy of Sciences and Arts and was located in St. Petersburg.) There he studied with Levich, the scientist best known for developing the rotating-disk electrode for use in researching electrochemical kinetics (the study of the rates at which chemical reactions occur).

Life's Work

Chizmadzhev's initial work in electrochemistry dealt with diffusion kinetics, the study of the rate at which ions (electrically charged atoms) move; this work involved studying pores, the tiny holes that allow substances to enter the skin. Working with a team under Frumkin, who is widely considered the father of electrochemistry, Chizmadzhev conducted research that was crucial to the development of the fuel cells that were used in early space flights. (He wrote about this work in his 1971 publication "Macrokinetics of Processes in Porous Media.")

Soon, Chizmadzhev began studying biophysics, the field that applies the laws and methods of physics to biological questions; Chizmadzhev specifically involved himself with studying axons (long, thin extensions of nerve cells, which transport information to target cells) and nerve systems. He also studied the transport of ions through the biomolecular lipid layers that comprise biological membranes, research that was useful in discovering how antibiotics are delivered to their target cells. Chizmadzhev next worked under Dogonadze, considered a pioneering figure, in the field of quantum electrochemistry, which takes aspects of quantum mechanics, a discipline that examines matter from the perspective that it is changing and unpredictable, and applied it to electrochemistry.

At the Frumkin Institute, Chizmadzhev's research has focused on membrane fusion and ionic transport. Membranes are cellular walls made of lipids, or fatty compounds. When two separate lipid membranes merge into a single bilayer, the process is known as membrane fusion, which is one of the fundamental chemical processes of life and is crucial to many biological processes, including cell fusion, endocytosis (in which substances enter a cell by way of a saclike vesicle), and the penetration of viruses into cells. One of Chizmadzhev's goals was to discover methods of increasing the permeability of skin tissue to allow for easier diffusion of drugs through the skin. He is credited with the first theoretical description of electroporation, a process in which cell walls are made more permeable using high-voltage current so new genetic material can be inserted. Electroporation is of particular interest to pharmaceutical companies; in 1992 Cygnus Therapeutic Systems—makers of transdermal patches for hormone therapy, smoking cessation, and contraception—entered into a sponsorship agreement with the Frumkin Institute, whose scientists have written more than fifty papers since the late 1970s on the topic. Cygnus hoped that electroporation would one day aid in the transdermal transmission of larger therapeutic molecules, such as proteins and peptides (molecules made up of two or more amino acids), and facilitate che-

> "The growth of fusion pores is poorly understood. Fusion proteins play a major role in creating a pore, but their importance to pore expansion is unclear."

motherapy treatment of dense tumors. In recent years the technology has seen increasing medical application, especially by oncologists. It is predicted that it will also be of great use as the field of gene therapy develops.

In "Dynamics of Fusion Pores Connecting Membranes of Different Tensions," published in the *Biophysical Journal* (May 2000), Chizmadzhev and his coauthors wrote, "The growth of fusion pores is poorly understood. Fusion proteins play a major role in creating a pore, but their importance to pore expansion is unclear." The paper comes to the conclusion that fusion pores tend to be in the shape of a toroid, a shape reminiscent of a doughnut, and the authors wrote, "By deriving rigorous equations for the energies of toroidal pores as a function of the membrane tension, we have been able to obtain a physically realistic framework for conceptualizing how biological fusion pores can enlarge. The biological process would not be obvious without quantitative exposition, but once the equations have been obtained, the importance of membrane mechanics in controlling the process of pore growth becomes apparent and logical."

Chizmadzhev has published numerous such papers, which, while not readily accessible to the layperson, have been considered of great importance in the scientific community and to the field of electrochemistry. Chizmadzhev, who has taught at the Moscow State University and the Moscow Institute of Engineering Physics for more than two decades, is known for his mentorship of numerous young scientists. Many of his former students have left Russia for the United States and Europe but remain in close contact with him. He has also been noted for his good humor and eagerness to share credit for his accomplishments with his colleagues. He has lectured at universities in Italy, Switzerland, and France, among other places.

In December 1987, Chizmadzhev was elected a corresponding member of the Russian Academy of Sciences. Since the late 1980s he has been the head of the editorial board for the journal *Biological Membranes*. He and his colleagues at the Frumkin Institute were awarded a State Prize, Russia's highest civilian honor, for their work in the field of electrochemistry.

Further Reading

Chizamdzhev, Yuri A., Kuzmin, Peter I, Weaver, James C. and Russel O Potts. "Skin Appendageal Macropores as Possible Pathway for Electrical Current." *Journal of Investigative Dermatology Symposium Proceedings* 3 (1998): 148–152. Print.

Grafov, Damaskin, and Pleskov Krishtalik. "Chronicle: Yuri Aleksandrovich Chizmadzhev." *Russian Journal of Electrochemistry.* 37.12 (2001): 1320–1321. Print.

Schuman, Tomas. *Love Letter to America*. Los Angeles: NATA, 1984. Print.

Ciechanover, Aaron

Israeli biochemist

Born: October 1947; Haifa, Israel

In 1976, after finishing his required service in the Israeli military, Aaron Ciechanover was faced with the choice of starting a residency—the step before becoming a surgeon—or going to graduate school to begin a career in research. "It was clear to me that I was heading for graduate school," Ciechanover wrote in his 2004 autobiography for the official Nobel Prize website. "My disillusionment from clinical medicine that disease can be cured based on understanding their pathogenetic mechanisms, along with a magical and enchanting attraction to biochemistry made the decision easy." On October 6, 2004, the Royal Swedish Academy of Sciences awarded Aaron Ciechanover, Avram Hershko, and Irwin Rose the Nobel Prize in Chemistry "for the discovery of ubiquitin-mediated protein degradation," the Nobel Foundation website stated. "Thanks to the work of the three Laureates it is now possible to understand at molecular level how the cell controls a number of central processes by breaking down certain proteins and not others."

Early Life and Education

Of Polish-Jewish parentage, Aaron (Aharonchik) Ciechanover (Checha) was born in the coastal city of Haifa, in what is now Israel, in October 1947, during the tumultuous year preceding the nation's founding. He was brought up in the working-class neighborhood of Hadar Hacarmel. His father, Yitzhak, was a law clerk who later became a lawyer; his mother, the former Bluma Lubashevsky, was a housewife and English teacher. By the time Ciechanover was four years old, his brother, fourteen years his senior, was already serving his compulsory time in the Israeli military. Ciechanover and his brother received "a liberal modern orthodox education," he wrote in his autobiography for the official Nobel Prize website. "It was extremely important for my parents to educate us as a new breed of proud Israeli Jews in their own independent country," he wrote. "My father inherited me with his love of Jewish studies and cultural life. To this very day, along with several physicians and scientists colleagues, I take regular lessons taught by a rabbinical scholar, on how the Jewish law views moral and ethical problems related to modern medicine and science."

In order to ensure that their children received a well-rounded education, Ciechanover's parents made sure their household was filled with books and classical-music recordings. He showed an early love for biology, collecting flowers and drying them between the pages of his brother's Talmud. When he was eleven years old, he received his first microscope—a gift his brother brought back from a trip to England. "With this microscope I discovered cells (in the thin onion epithelium) and did my first experiment in osmosis, when I followed the alteration in the volume of the cells after immersing the epithelium in salt solutions of different strengths," Ciechanover wrote in his Nobel autobiography.

Ciechanover's mother died in 1958, while he was in elementary school. His father raised him for several years before he, too, died, in 1964. Having lost both of his parents, he was left to the care of his aunt Miriam, his brother, and his sister-in-law, Atara. Their help and guidance, he recalled in his autobiography, enabled him "to seamlessly complete my high school studies in the same class and along with my friends—without interruption. The other option was to move to Tel Aviv, to my brother's home, but this would have been much more complicated. So I spent the weekdays with my aunt in Haifa, studying, and the weekends and holidays with my brother and sister-in-law, in Tel Aviv. Their help was a true miracle, as thinking of it retrospectively, being left alone without parents at the age of sixteen, the distance to youth delinquency was shorter than the one to the high school class. Yet, with the help of these wonderful family members, I managed to continue."

In high school Ciechanover opted to major in biology. In trying to decide on a career, he discounted physics and chemistry as "strong mechanistic disciplines built on solid mathematical foundations," he wrote in his autobiography. He preferred biology, with its many secrets still unknown to scientists, but he was "afraid to get lost." A good compromise, he felt, might be medicine. In addition to finding medicine to be a good mix of all the sciences—not to mention a well-paying profession parents often encouraged their children to pursue—Ciechanover liked the idea of curing diseases. His mind was made up when he learned that the Israeli government allowed medical students to postpone military service.

In 1965 Ciechanover enrolled at Hebrew University, in Jerusalem, then the only medical school in the country. In his fourth year at the school, he decided to become a research doctor rather than a practicing physician. "I felt restless and started to realize how little we know, how descriptive is our understanding of disease mechanisms and pathology, and as a consequence how most treatments are symptomatic in nature rather than causative," he wrote in his autobiography. "The statement 'with God's help' that I heard so frequently from patients that were praying for cure and health, took on a real meaning. I had a feeling clinical medicine was going to bore me, and decided to take one year off in order to 'taste' true and 'wet' basic research." With his brother's help, he postponed military service another year and enrolled in a research program that was studying the causes of fatty liver in rats.

From that year of laboratory work, Ciechanover realized biochemistry was his calling. After finishing this research, he completed his required internship at the Faculty of Medicine at the Technion-Israel Institute of Technology, in Haifa. The move allowed him to study under Avram Hershko, a young biochemist who had recently done his post-graduate studies in San Francisco, California. Ciechanover earned his MD in 1973 and served in the army for the next three years. After serving as the physician on a missile boat, he moved to the research and development unit of the Medical Corps, where he designed "sophisticated devices for the soldier in the battlefield," according to his autobiographical statement. "Because of the broad range of experiences acquired, the military service has been my ever best school for real life 'sciences,'" Ciechanover wrote. Following his discharge, in 1976, he made what he called in his autobiography "the most important decision" of his career and chose scientific research over clinical medicine. He returned to Technion, where he earned his doctorate in 1981.

Life's Work

At the Technion-Israel Institute, under the tutelage of Avram Hershko, Ciechanover began researching protein degradation—the process through which an organic cell breaks down proteins—a field that would eventually lead to the Nobel Prize. Speaking to Joanna Rose in an interview for the Nobel Foundation website (October 6, 2004), Ciechanover explained that "when I was a graduate student with Avram Hershko, the other laureate, and then with Ernie [Irwin] Rose, we saw something that wasn't in the main focus of science. It was protein degradation. Everybody looked into protein synthesis . . . into DNA and how the genome is being translated into the proteome. And we said, 'No. Maybe there is something on the other side.'"

Hershko and Ciechanover first set out to analyze why the degradation of proteins within a cell required energy in the form of adenosine triphosphate (ATP). Hershko "wondered if the requirement for ATP hinted at a novel system of protein degradation," as Eugene Russo wrote in the *Scientist* (October 16, 2000). Through a series of experiments conducted in 1978, Hershko and Ciechanover were able to isolate the substance that catalyzed the process of protein degradation. Working with young blood cells, Ciechanover and Hershko first removed the hemogoblin—a substance that had previously skewed their results—from the samples. The emerging extract, they subsequently realized, could be separated into two parts, both of which were inactive on their own but when combined caused ATP-dependent protein degradation to commence. From this data Ciechanover and Hershko determined that one of the active components was a protein called ATP-dependent proteolysis factor 1, which they gave the acronym APF-1. Over the next several years, the two men, working in collaboration with Irwin Rose at the Fox Chase Cancer Center in Philadelphia, Pennsylvania, deciphered the mode of action of APF-1 in protein degradation. In 1980 Ciechanover, Rose, and Hershko, along with Hannah Heller, published two seminal papers highlighting their discoveries: the first showed that APF-1 formed a stable bond with a variety of proteins in liver cell samples; in the second, they elucidated the process through which "many APF-1 molecules could be bound to the same target protein," the Nobel Foundation website reported. Meanwhile, the research of Keith D. Wilkinson, Arthur L. Haas, and Michael K. Urban showed that APF-1 was ubiquitin, a previously known protein of unknown function. It was known, however, that ubiquitin binds once to another protein called histone; therefore,

> **"My disillusionment from clinical medicine that disease can be cured based on understanding their pathogenetic mechanisms, along with a magical and enchanting attraction to biochemistry made the decision easy."**

the fields of histone monoubiquitination and protein polyubiquitination now converged. The process of polyubiquitination is the so-called "kiss of death" whereby a protein is marked for degradation. "At a stroke," the Nobel Foundation website observed, "these entirely unanticipated discoveries changed the conditions for future work: it now became possible to concentrate on identifying the enzyme system that binds ubiquitin to its target proteins." Since ubiquitin, derived from the word ubiquitous, is so common in a variety of organisms, it was theorized that the substance and the process it spurred had a vast impact on a cellular level. Moreover, Ciechanover and Hershko hypothesized that since the energy for the polyubiquitination derived from ATP, the amount of ATP involved in a particular reaction would allow the "cell to control the specificity of the process." In other words, the level of ubiquitin determined the nature of the subsequent reaction.

The Multistep Ubiquitin Tagging Hypothesis

Having established the fundamental components of polyubiquitination, Ciechanover and Hershko posited what the Nobel Foundation website referred to as the "multistep ubiquitin tagging hypothesis." Between 1981 and 1983 they identified three different enzyme activities that can occur during polyubiquitination, which they named E1, E2, and E3. The E1 enzyme activates the ubiquitin molecule, which is then forwarded to an E2 enzyme; in turn, the Nobel Foundation website explained, "the E3 enzyme can recognize the protein target which is to be destroyed. The E2-ubiquitin complex binds so near to the protein target that the actual ubiqutin label can be transferred from E2 to the target." The ubiquitin is then transferred to the substrate, a process mediated by the E3 enzyme and repeated until a connected chain of ubiquitin molecules emerges. Then the proteasome, a barrel-shaped component of a cell that breaks up or degrades proteins within the cell, recognizes the chain, which is subsequently disconnected and then cut up in the proteasome. (The proteasome was discovered later. Its core catalytic part was identified by Sherwin Wilk and Marian Orlowski and its function in ubiquitin-conjugates degradation by Martin Rechsteiner and then Alfred Goldberg.)

For medical purposes, this new understanding of the process of protein degradation had a variety of implications. Many diseases—cystic fibrosis, Parkinson's disease, and certain forms of cancer, for instance—are caused by the malfunctioning of enzymes that are supposed to help bond ubiquitin to or detach it from its protein targets. Hershko explained to Judy Siegel for the *Jerusalem Post* (October 3, 2003), "If proteins are not degraded at the right time, the cell continues to divide unchecked. This is what happens in many cancer cells; something has gone wrong in the ubiquitin system so there is no control over cell division." Consequently, scientists are now beginning to design medicines that protect against the degradation of certain proteins or promote the destruction of them. One such medication, Velcade, a potential treatment for multiple myeloma that is based on inhibition of the proteasome, is already being tested, and many more will soon undergo clinical trials. Reflecting on the decision to study protein degradation, Ciechanover has said that the field seemed important to him and the two men with whom he shared the Nobel Prize, though they never imagined that it would be quite as important as it turned out to be. As most others did not regard protein degradation as an important field of research—and those that did, including experts in the area of proteolysis, did not believe Ciechanover and his colleagues' findings—Ciechanover and his colleagues had no competition for many years.

Since the discoveries in the late 1970s and early 1980s, Ciechanover has continued his work in ubiquitin-mediated protein degradation. He carried out his postdoctoral fellowship work at the Massachusetts Institute of Technology (MIT) in the laboratory of Harvey Lodish, where, in collaboration with Alexander Varshavsky (who studied ubiquitin modification of histones) and Daniel Finley, they identified the first living cell mutant of the system. Ciechanover became an associate professor at the Technion-Israel Institute of Technology's Unit of Biochemistry in 1987 and a full professor in 1992. He has been a visiting professor of pediatrics at the Washington University School of Medicine in St. Louis, Missouri. His current work focuses on the degradation of nuclear oncoproteins, which are proteins that control cell division. Mutated or absent oncoproteins are associated with certain forms of cancer.

While there had been speculation for several years that Hershko, Rose, and Ciechanover might be in line for a Nobel Prize, Ciechanover had tried to avoid thinking about the possibility. He believed, however, that if he ever did win the award it would be in medicine, not chemistry. When the Nobel committee announced, on October 4, 2004, that the prize in medicine would go to Richard Axel and Linda B. Buck, Ciechanover's students walked disappointedly to his office, though Ciechanover rejected the idea of even discussing the issue with them. The announcement that he, too, would receive the award came two days later.

Since winning the Nobel Prize, Ciechanover has used his new prominence to strongly encourage the Israeli government to increase its funding for education: "We are destroying with our own hands the essential foundation of the State of Israel," he told Traubman. "We do not have the correct national priorities. We are a country that lacks riches. The Jewish brain, that's what we have. Everything that we've had and will have in this country is the direct and clear product of higher education. If we harm this system, we will drastically decline and cease to exist." He also pointed out to Traubman that the lack of government support for his projects had forced him to constantly scramble for grant money and pursue other forms of income: "So I moonlight. I sit on all kinds of scientific councils of biotechnology companies. Do I like this? If I could, I wouldn't do this." Ciechanover has received offers of employment from major laboratories in the United States. Regarding the choices facing him in light of those opportunities, he told Traubman, "What will keep me here? My Israeliness." Continuing that conversation with Noble Prize winners, Ciechanover stated in an email message that he remains in Israel "in spite of the conditions and not because of them."

In addition to the Nobel Prize, Ciechanover has received other awards and accolades over the years. In 1997 the Technion-Israel Institute honored him with its Henry Taube Prize; in 1999 he received the Wachter Prize from the nation of Austria, an honor he shared with his former teacher Hershko. In addition, Ciechanover was honored, along with his former collaborator Varshavsky and, again, Hershko, with the 2000 Lasker Award for Basic Medical Research, a prize often considered the harbinger of a future Nobel. In 2007 Protalix Biotherapuetics Inc., a pharmaceutical company, named him to its scientific advisory board.

Though a renowned scientist, Ciechanover fosters an informal atmosphere at his lab, in part by wearing jeans and casual shirts. While he is considered friendly and candid, Ciechanover has also earned a reputation for prickliness, an attribute he is trying to curtail now that he is a Nobel laureate: "I have a very sharp tongue and I'm sure that I'll still slip up," he told Traubman. His taste in music is broad, ranging from the Swedish supergroup Abba to classical. He has sat on the board of directors of the Haifa Theatre. In his leisure time Ciechanover enjoys walking along the sea with friends and participating in a club for enthusiasts of old clocks. He and his wife, Menucha, a graduate of the Harvard School of Public Health and the chairman of the department of geriatrics at an Israeli hospital, have a son, Isaac, who is currently a student of social work at Haifa University.

Further Reading

Ciechanover, A., Hod, Y. and Hershko, A. (1978). "A Heat-stable Polypeptide Component of an ATP-dependent Proteolytic System from Reticulocytes." *Biochem. Biophys. Res. Commun.* 81 (1978): 1100–1105. Print.

Hershko, A., and Aaron Ciechanover. "Mechanisms of intracellular protein breakdown." *Annu. Rev. Biochem.* 51 (1982): 335–364. Print.

Ciechanover, Aaron. "Proteolysis: from the lysosome to ubiquitin and the proteasome." *Nat. Rev. Mol. Cell Biol.* 6 (January 2005): 79–87. Print.

"Profile: Aaron Ciechanover." *Agence France-Presse* 1 January 2001. Print.

Rolnik, Guy, and Oren Majar. "A meeting of Minds on Matter, and Spirit." *Haaretz* 19 July 2011: AV 22. Print.

Djerassi, Carl

American chemist

Born: October 29, 1923; Vienna, Austria

In 1999, when the *London Times* compiled its list of the thirty most important people of the millennium, many familiar names were included: Isaac Newton, Galileo, William Shakespeare, Ferdinand Magellan, Mozart, Charles Darwin, Louis Pasteur, and Albert Einstein all got their due. One name, however, was unfamiliar to many: that of the organic chemist Carl Djerassi. Known as the father of the Pill, the first oral contraceptive, Djerassi has been a controversial figure since his invention took the world by storm in the 1960s and forever changed society's view of sex. Djerassi's contributions to science extend beyond that groundbreaking achievement. His name is on the patent for the first antihistamines, and following his work on the Pill, he was a pioneer in the development of pest-control substances. Over the last decade Djerassi has turned to writing fiction. He has published several books of "science-in-fiction," a term he coined himself; in those works, he has explored the fallibility of scientists and the societal conflicts raised through their quests for recognition. He is also the author of two plays as well as collections of essays and poetry.

Early Life and Education

Carl Djerassi was born on October 29, 1923, in Vienna, Austria, the son of Jewish parents—Samuel and Alice Djerassi—who were both physicians. His rearing was largely nonreligious. He spent the first years of his childhood in Bulgaria, his father's native country, returning to Vienna to start school. Samuel Djerassi specialized in treating venereal disease in the days before the advent of penicillin; Carl Djerassi's bar mitzvah was postponed when a wealthy syphilis patient arrived at his father's clinic.

It was around this time, when Djerassi was in his early teens, that he discovered his parents had divorced when he was six. His father had stayed in Bulgaria, seeing the boy during his visits to Vienna or when Carl visited him in the summers. "I suppose I was simply too young and generally too happy to wonder that my parents didn't live together," Djerassi told Nicholas Wroe for the *London Guardian* (August 26, 2000). He found out the truth when his father introduced him to his girlfriend.

In 1938, during the *anschluss*—the incorporating of Austria into Nazi Germany—Djerassi's father returned to Vienna and remarried his mother in order to take the family out of the country; once they were safely away from Austria, thanks to Samuel Djerassi's foreign nationality, the marriage was annulled. Djerassi's mother then traveled to England to arrange her and her son's entry into the United States. Carl Djerassi, meanwhile, remained in Bulgaria, where he learned English.

Djerassi and his mother arrived, with next to no money, in the United States in 1939. Realizing that an education was of the utmost importance for his future, the teenage Djerassi wrote a letter to Eleanor Roosevelt, the wife of President Franklin Delano Roosevelt, explaining his situation and asking for financial assistance. Eleanor Roosevelt passed Djerassi's letter on to the Institute of International Education. As a result, Djerassi was offered a scholarship to a Presbyterian college in Missouri. He completed his chemistry degree at Kenyon College, in Gambier, Ohio, at age nineteen, then pursued postgraduate studies at the University of Wisconsin. During that time he married his first wife, Virginia.

Life's Work

Djerassi became a naturalized US citizen in 1945. Shortly thereafter he moved to New Jersey to work for CIBA, a Swiss pharmaceutical company. In 1949 he began working for Syntex S.A., in Mexico City, Mexico, where he started out by studying the uses of cortisone. Soon he was researching the treatment of certain types of menstrual disorders as well as cancer. In 1951, by synthesizing different hormones, Djerassi—along with Min Chuch Chang, Gregory Pincus, and John Rock—formed the foundation of what would become known as the Pill, the first synthetic contraceptive that could be taken orally. The Pill was made available to the public in the early 1960s, and, by divorcing sex and pregnancy, inaugurated a period of great social change.

Following the advent of the Pill, Djerassi became something of a celebrity. In 1973 he was awarded the National Medal of Science for his work. He remained modest about his discovery, however. "Yes I am proud to be called the father of the Pill," he told Nicholas Wroe. "But identifying scientists is really only a surrogate for identifying the inventions or discoveries. Maybe it is true that Shakespeare's plays would never been written if it wasn't for Shakespeare. But I'm certain that if we didn't do our work then someone else would have come along shortly afterwards and done it."

> "I think that we as scientists should educate the public about the scientific and technological advances so that society can decide how to best use them. This is my missionary obsession."

In 1968 Djerassi founded his own company, Zoecon, which developed new approaches to insect control. As the decade closed he came out openly against the United States' involvement in the Vietnam War. Several years later it was revealed that, because of his involvement in the Democrat George McGovern's 1972 presidential campaign, Djerassi had been placed on the infamous enemies list of Republican president Richard Nixon.

Since the invention of the Pill, Djerassi has spent a great deal of time discussing—and facing controversy over—its possible side effects. The main accusation is that the Pill increases the risk of blood clots in women; some have linked the Pill to an increased risk of cancer and to excessive bleeding during menstruation. Djerassi has dismissed those claims, pointing to studies that place the number of blood clots in users of the Pill at 15 to 25 per 100,000, compared with 60 per 100,000 for pregnant women; he also asserts that links between the Pill and cancer or excessive bleeding during menstruation have never been proven. Djerassi is an outspoken proponent of oral contraceptives' being made available without prescription and is critical of the government's decreased funding for the further development of the Pill.

Today, Djerassi is an accomplished author with several novels, memoirs, and plays to his credit. He writes "science-in-fiction," which he has defined as fictional writing based on science, often discussing the social and personal dilemmas brought about by achievements in that field. He believes that discussions of his fiction can also serve as forums for examining the moral and ethical implications of scientific advances. "The more different ways we have of teaching certain difficult things the better off we are. I think that we as scientists should educate the public about the scientific and technological advances so that society can decide how to best use them. This is my missionary obsession," he told Nicholas Wroe. "Just because you have won a Nobel Prize doesn't mean you're qualified to make these important judgments. Ethics is a social construct and is something that society should establish. But in order to make educated decisions about this you must be aware of it."

Books and Publications

In Djerassi's first novel, *Cantor's Dilemma* (1989), a well-known biologist develops a theory to explain why cells become cancerous; he knows that his theory can win him the Nobel Prize. When he finds out that his lab assistant may have falsified some of the lab results, he attempts to cover up the inconsistencies so that the prize can be his. Djerassi followed up his 1990 nonfiction work *Steroids Made It Possible* with *The Pill, Pygmy Chimps, and Degas' Horse: The Remarkable Autobiography of the Award-Winning Scientist Who Synthesized the Birth Control Pill* (1972). Next came his second science-in-fiction book, *The Bourbaki Gambit* (1994), which involves four scientists who are forced into retirement but decide to continue their research under a collective pseudonym. The group then makes a significant discovery that threatens to tear them apart. The book *Marx, Deceased* (1996) is about a novelist, Stephen Marx, who, wondering how he will be remembered, stages his own death. He begins writing under a new name, then becomes romantically involved with a journalist who has uncovered his deed. Meanwhile, his wife, who believes she is a widow, begins a relationship with a critic who is writing an evaluation of Marx's career. The critic harbors a grudge against Marx because the novelist had an extramarital affair with his wife.

In 2001 Djerassi published *This Man's Pill—Reflections on the 50th Birthday of the Pill*, in which he discussed common criticisms of the oral contraceptive—from religious fundamentalists who blame its invention for the general moral decline of mankind, and from feminists who see the Pill as the ultimate manifestation of the patriarchal nature of society, since it was developed to be used exclusively by women. The bulk of the book explores how the Pill affected Djerassi himself, forcing him to explore social ramifications he had not initially considered. ("Until 1969, I would have described myself as a 'hard' scientist, the proudly macho adjective employed by chemists and other physical scientists to distinguish their work from the 'soft,' fuzzy fields such as sociology or even psychology," Djerassi wrote in the book.)

Oxygen (2001) is a play cowritten by Djerassi and Roald Hoffmann. Its action alternates between the years 2001 and 1777. In the present day, the Nobel Foundation decides to inaugurate a "Retro-Nobel" Award for important discoveries that preceded the creation of the Nobel Prize. One such discovery is that of oxygen, which raises the question of who should be credited. The setting then shifts to 1777, when three men vied for that distinction. The play premiered at the San Diego Repertory Theatre on April 2, 2001.

The subject of Djerassi's play *An Immaculate Misconception* (1998) is assisted reproductive technologies (ART), the process by which an egg can be successfully fertilized with the injection of a single sperm under a microscope, followed by the insertion of the egg into a woman's uterus. Djerassi told Norman Swan for the *Health Report* (November 20, 2000), "Sex of course still continues to be done in the usual way and that's for reasons that we always do: love, lust, pleasure. But reproduction separately, fertilization, under the microscope, so to speak, . . . this separation of sex and reproduction . . . was not possible before, so therefore implicit in every sexual act was the fact that you might get pregnant, and that of course is you might say the Catholic dictum: you should not have sex for fun, you should have sex with at least the possibility, you can do it for fun but only with all the possibility of reproduction. Well that stopped in a way you could say in 1960, with the introduction of oral contraceptives and IUDs, because these are the two methods of contraception that separated sex from conception." Djerassi believes that the future of birth control lies with ART. He foresees a future in which young men will deposit sperm in banks for later artificial insemination, then undergo vasectomies, making sex and reproduction two entirely separate endeavors. "I'm trying to smuggle important scientific ideas out to the general public. The vast majority of legislators haven't got the foggiest idea about science or technology, and yet many of the decisions they make are based on these things," he told Martha J. Heil for *Discover* (June 2000).

Djerassi has received eighteen honorary doctorates in addition to many other honors, including the first Wolf Prize in Chemistry, the first award for the Industrial Application of Science from the National Academy of Sciences, and the American Chemical Society's highest award, the Priestley Medal. In 2011, he was awarded the prestigious Edinburgh Medal. He is one of only a few scientists to receive both the National Medal of Science and the National Medal of Technology. He is the author of more than 1,200 scientific publications and seven monographs. In 2010, Djerassi was elected a foreign member of the Royal Society of London, a fellowship of more than one thousand distinguished scientists.

Personal Life

Djerassi divorced his first wife in 1950 and his second wife, Norma, in 1976. He was married again on June 21, 1985; his third wife, Diane Middlebrook, is a writer and professor at Stanford University. Djerassi's daughter, Pamela, was born in 1950; she committed suicide at age twenty-eight. Djerassi founded the Djerassi Resident Artists Program (an artists' colony on his California ranch) in her memory in 1979. The program has benefitted more than one thousand writers, painters, choreographers, and sculptors. He has one son, Dale, from his second marriage.

Further Reading

Djerassi, Carl. This Man's Pill: Reflections on the 50th Birthday of the Pill (Popular Science). New York: Oxford UP, 2004. Print.

---. *Cantor's Dilemma*. New York: Doubleday, 1989. Print.

---. *The Pill, Pygmy Chimps, and Degas' Horse: Autobiography of Carl Djerassi*. New York: Basic, 1992. Print.

Williams, Zoe. "How the Inventor of the Pill Changed the World for Women." *Guardian*, 29 October 2012. Print.

Doudna, Jennifer

American biochemist

Born: February 19, 1964; Hawaii

Jennifer Doudna, a professor of biochemistry and molecular biology at the University of California, Berkeley, has done groundbreaking research into the nature and function of ribonucleic acid, known as RNA. In the 1980s scientists discovered that RNA, previously thought to be simply a passive carrier of genetic material (a determinant of, among other traits, the color of one's eyes and one's height), also functioned at times like an enzyme. (Enzymes act as organic catalysts, speeding up various biochemical reactions in an organism. Enzymes are at work, for example, in human digestion and in the ripening of fruit.) Doudna is investigating the structure of ribozymes—RNA with enzymatic properties—and how they affect biochemical processes, including the spread of damaged or viral cells, as happens in cases of cancer and AIDS. Her RNA research, which has been characterized as among the most important in her field in recent years, may help scientists develop treatments for those diseases. While the possible applications could be of enormous importance, Doudna told Melissa Marino in an article for the *Proceedings of the National Academy of Sciences* (December 7, 2004) that her interest in RNA is more basic. "For me, the bigger question is, 'Can we get enough information so that we can understand the chemical basis for RNA's many biological functions?'" she said. "It will be exciting to make meaningful comparisons between the chemistry of ribozyme reactions and what happens in protein enzymes that carry out similar reactions." Scientists have hypothesized that early life forms may have been based largely on RNA (or a related molecule), and if Doudna discovers many similarities between the structures and functions of RNA and proteins, that hypothesis will gain credence. "Obviously, until we build a time machine, we can't really go back and look at that," Doudna told Marino; thus, such an idea can never truly be tested. But, she added, "I think the idea that RNA might have played a critical role" in the development of early life forms "is very tantalizing."

Early Life and Education

Jennifer Doudna was born on February 19, 1964, and raised in Hawaii. Her parents both worked in academia. Although their professional focus was on the humanities, they were interested in astronomy and geology, among other subjects, and encouraged Doudna to pursue her own passions. In high school Doudna took her first chemistry class. She became fully hooked on science after reading *The Double Helix: A Personal Account of the Discovery of the Structure of DNA* (1968), by James D. Watson, who had done his seminal work with Francis Crick in the 1950s. Doudna's parents were friends of Don Hemmes, a mycologist and professor of biology at the University of Hawaii at Hilo, and they arranged for her to work in Hemmes's lab for a summer, studying mushrooms. After high school Doudna enrolled at Pomona College, in California, where she studied French and medieval history in addition to science. At Pomona she worked in the chemistry lab of Sharon Panasenko, who was her faculty adviser. Doudna has described Panasenko as being a strong female role model—a rarity in the male-dominated sciences. Doudna earned a bachelor's degree in chemistry in 1985 and then entered Harvard University in Cambridge, Massachusetts, to pursue a doctoral degree in biochemistry, which she earned in 1989. At Harvard Doudna studied with Jack Szostak, who was doing exciting research on RNA. Doudna remained at Harvard as a postdoctoral fellow until 1991. While in Cambridge Doudna also served as a research fellow at Massachusetts General Hospital and Harvard Medical School.

After she left Harvard, in 1991, Doudna continued her RNA research as a Lucille P. Markey scholar and postdoctoral fellow at the University of Colorado. There she worked in the laboratory of the biochemist Thomas Cech, a pioneer in the study of RNA. Cech, along with the biochemist Sidney Altman, had documented the existence of ribozymes in the early 1980s. Their research, which won them the Nobel Prize in chemistry in 1989, proved that RNA was not merely a passive transportation device for an organism's genetic code, as had previously been believed, but also acted as an enzyme that stimulated chemical reactions leading to the growth of new cells. In Cech's lab Doudna began working on mapping the three-dimensional chemical

> **"I think the idea that RNA might have played a critical role [in the development of early life forms] is very tantalizing."**

structure of a portion of the *Tetrahymena thermophila* ribozyme. (*Tetrahymena thermophila* is the single-celled pond organism that Cech and Altman had used earlier to prove the existence of ribozymes.)

Life's Work

In 1994 Doudna joined the faculty of Yale University in New Haven, Connecticut, as a professor of biochemistry and biophysics. At Yale Doudna continued to try to better understand the molecular structure of RNA. Her research was funded in part by the National Institutes of Health and various federal grants. Cech continued to collaborate with Doudna as well as with her Yale colleagues. The first task facing them was to grow in crystal form the *Tetrahymena thermophila* ribozymes they had isolated, so that they could examine the ribozymes' atomic structure. To grow the crystals, the researchers used salt compounds, which counteracted the ribozyme molecules' tendency to repel one another rather than crystallize. It was a difficult and time-consuming process. Once Doudna and her colleagues had grown a ribozyme crystal that was large enough, they placed it in an X-ray crystallography machine. The X-rays the machine fired at the crystals produced reflections of the ribozymes' structure, which Doudna and her team studied and copied; eventually, in about 1996, they were able to build a three-dimensional model of a portion of the ribozyme's molecular structure. Though Doudna and her colleagues had mapped only about half of their subject, the model they created represented a tremendous advance. The structural model "showed us how RNA is able to pack helices together to form a three-dimensional shape," Doudna explained to Marino, "which is much more reminiscent of what we see in proteins than anybody had been previously aware of for RNA." In further studies Doudna and her colleagues found that a portion of the ribozyme folded in upon itself, a phenomenon similar to that observed in proteins. For this work Doudna won the Johnson Foundation Prize for Innovative Research, the Beckman Young Investigator Award, the Searle Scholar Award, and the David and Lucille Packard Foundation Fellow Award.

On October 8, 1998, Doudna and her collaborators reported that they had found an RNA enzyme in a human pathogen in the journal *Nature*—a scientific first. In particular, they described the crystal structure of a ribozyme that contributes to the replication of the hepatitis delta virus, which can cause a potentially fatal secondary infection; such infections occur most frequently among people living in developing nations. On October 9, 1998, in an issue of *Science*, Cech described the three-dimensional structure of the *Tetrahymena thermophila* ribozyme. Later in 1998 Doudna was granted tenure at Yale. (Tenure affords a professor greater respect and, in practical terms, means that he or she cannot be fired summarily.) She was named Henry Ford II Professor of Molecular Biophysics and Biochemistry in 2000.

In 2003 Doudna joined the faculty of the University of California, Berkeley as a professor of biochemistry and molecular biology. The move allowed her, her husband, and their son to be closer to members of their extended family. It also gave Doudna easy access to the synchrotron (an apparatus that imparts high speed to charged particles using magnetic and electric fields of varying frequencies) at the Lawrence Berkeley National Laboratory. Much of her most recent work concerns viral RNA and its functions, in particular a section of viral RNA called the internal ribosome entry site (IRES). (Ribosomes, a component of cells, synthesize proteins.) Doudna described IRES to Marino as a "pretty amazing structure that's basically able to grab the ribosomes of infected cells and hijack them for making viral proteins." She continued, "I think this is the project in the lab that has the greatest potential to lead to something that might have an impact on human health." The practical applications of her work include the creation of ribozyme-based drug therapies to stop the spread of mutated cells. Pharmaceutical companies began putting Doudna's structural model to use soon after she and her team created it. As reported by David

Fletcher for the *Yale Daily News* (September 26, 1996), researchers working for such companies as Innovir (which was founded by Sydney Altman) sought to use Doudna's model to learn how to manipulate cells affected by various genetic diseases, including cystic fibrosis and sickle-cell anemia.

Doudna is a member of the editorial boards of *Current Biology* and the *Journal of Molecular Biology*. She has authored or coauthored more than seventy monographs and papers on her research. In 2000 Doudna became the third woman ever to be honored with the prestigious Alan T. Waterman Award from the National Science Foundation; the award included a grant of $500,000 over the course of three years. Joan Steitz, one of Doudna's former colleagues in Yale's Biochemistry Department, who recommended Doudna for the award, told a reporter for *AScribe Newswire* (April 18, 2000) that Doudna possesses a "particular constellation of qualities as a scientist. She knows a lot of biochemistry. . . . And she has all sorts of guts and determination, and that's the reason why she tackled what was a very, very hard problem. . . . There can be no question that her pioneering accomplishments have changed the way the scientific community thinks about RNA molecules." In addition to her teaching responsibilities at the University of California, Berkeley, Doudna is an investigator at the Howard Hughes Medical Institute, in Chevy Chase, Maryland, a post she has held since 1997. She has been elected to the American Academy of Arts and Sciences and the National Academy of Sciences, honors usually reserved for older scientists with longer careers.

Further Reading

Berry, K. E., Waghray, S., Doudna, J. A. "The HCV IRES pseudoknot positions the initiation codon on the 40S ribosomal subunit." *RNA* 16.8 (2010): 1559–1569. Print.

"Jennifer A. Doudna, Ph.D." *HHMI Investigators*. Howard Hughes Medical Institute. Web. 13 August, 2012. <http://www.hhmi.org/research/investigators/doudna_bio.html>.

"Jennifer A. Doudna." *Molecular and Cell Biology*. University of California at Berkeley. Web. 13 August 2012. < http://mcb.berkeley.edu/index>.

Jinek, M., Coyle, S. M., Doudna, J. D."Coupled 5' nucleotide recognition and processivity in Xrn1-mediated mRNA decay." *Mol. Cell* 41.5(2011): 600–8. Print.

Marino, Melissa. "Biography of Jennifer A Doudna." *Proceedings of the National Academy of Sciences* (7 December 2004): 16987. Print.

Sashital, D. G., Jinek, M., Doudna, J. A. "An RNA-induced conformational change required for CRISPR RNA cleavage by the endoribonuclease Cse3." *Nat. Struct. Mol. Biol.* 18.6 (2011): 680–7. Print.

Good, Mary L.

American chemist

Born: June 20, 1931; Grapevine, Texas

A former undersecretary of the US Department of Commerce, a former senior vice president for technology at AlliedSignal Inc., and the holder of an endowed chair at the University of Arkansas, Mary L. Good has pursued successful careers in three separate fields: government, industry, and academia. Currently, she chairs the board of directors of the American Association for the Advancement of Science and is a managing member of Venture Capital Investors, a group of Arkansas business leaders who support technology-based enterprises. Good has earned many honors, among them the American Chemical Society's Priestly Medal, the highest award bestowed by that organization; the National Science Foundation's Distinguished Public Service Award; the American Association for the Advancement of Science Award; the Albert Fox Demers Medal from Rensselaer Polytechnic Institute; and the American Institute of Chemists' Gold Medal.

Early Life and Education

The daughter of John W. Lowe, a school superintendent, and Winnie Mercer Lowe, a schoolteacher, Good was born Mary Lowe in Grapevine, in the Dallas-Fort Worth area of Texas, on June 20, 1931. One of four children, she grew up in Arkansas. After she graduated from high school in Willisville, she enrolled at the Arkansas State Teachers College (now the University of Central Arkansas), in Conway. She has told interviewers that, unlike some of her colleagues, as a child she did not plan to become a scientist. "I never even had a chemistry course until college," she told Carol Kleiman for the *Chicago Tribune* (May 20, 1985). She also noted to Kleiman that her relatively late exposure to the subject "might have been for the better because some high school courses at that time weren't very good." When she entered college, Good intended to major in home economics, in large part because she considered it a useful subject, but she soon abandoned that plan. "In my freshman year," Good told Janice R. Long for *Chemical and Engineering News* (May 13, 1996), "I took my first chemistry course. My teacher was an elderly professor who was absolutely fantastic. He loved his students and he loved his subject. I had a ball. The upshot was that by the second semester I had changed my major to chemistry. I loved working in the lab, particularly wet chemistry, the qualitative analysis lab." In 1950, the nineteen-year-old Good earned a bachelor's degree in chemistry. She completed her undergraduate education in just three years by attending classes in the summer as well as during the fall and spring terms. She did so partly, as she explained to Long, because the four siblings in her family were all close in age, and she "needed to get through and out so the younger ones could go to school." Good won a fellowship from the University of Arkansas at Fayetteville to pursue graduate study in inorganic chemistry and radiochemistry. She completed a master's degree at that school in 1953 and earned a doctorate there in 1955. Meanwhile, in May 1952, she had married Billy Jewel Good, who was then a graduate student in physics. They have two children: Billy John, now a biologist, and James Patrick, an architect. "Fortunately," Good told Kleiman, "I was never told I couldn't have a career and a family. It was difficult at times, but Bill is very supportive."

Life's Work

Before she completed her graduate education, Good began teaching chemistry at Louisiana State University in Baton Rouge—a job she took in part because her husband wanted to complete his doctorate in physics, and Louisiana State, unlike the University of Arkansas, offered a doctoral program in that field. Her decision to pursue a career in academia was further influenced by the fact that she was a young mother. "I had a nine-month-old child at the time," she told Long, "and in those days an academic schedule offered more flexibility than an industry job would have." Good was promoted from instructor to assistant professor in 1956. Two years later she transferred to the New Orleans campus of Louisiana State University, where she was named associate professor. By 1963 Good had become a full professor, a position that she held for seventeen years. In 1974 she received an endowed chair and was named Boyd Professor of Chemistry. During her academic career Good published nearly one hundred research papers and a number of review articles. She also wrote the textbook

> "I was never told I couldn't have a career and a family. It was difficult at times, but Bill [Good's husband] is very supportive."

Integrated Laboratory Sequence: Volume III—Separations and Analysis (1970). "What I really enjoyed [was] working with the students," she told Long, "particularly my graduate students. There were close to twenty who worked with me over the years, and I keep in touch with many of them."

In 1980, Good left academia to become the director of research at UOP Inc., a branch of Signal Research Center Inc., an energy and technology conglomerate with annual revenues of $6 billion at that time. "I couldn't turn it down," she told Long of the UOP job offer. "It was so challenging and so interesting, I had to see if I could succeed.... At the time, I had pretty much accomplished all I wanted to at the university." Among her accomplishments at UOP, which focused on energy and petroleum research, was the synthesis of a chemical that discouraged barnacles from clinging to painted boats and docks. Good prospered at UOP, and in 1985 she was named president and director of research for Signal Research Center.

Signal underwent extensive restructuring in the wake of the buyouts and acquisitions that swept the chemical industry in the 1980s, and in 1985 it merged with the Allied Corp. to form AlliedSignal Research and Technology Laboratory, a $12 billion-a-year company that produced chemicals, fibers, plastics, and aerospace and automotive parts. In 1986 Good was named president of engineered materials research at AlliedSignal, and in 1988 she became senior vice president for technology, a post she held for five years. The year 1988 also saw the publication of the book *Biotechnology and Materials Science: Chemistry for the Future*, which she edited.

Meanwhile, Good had become increasingly involved with the public sector. In 1980 President Jimmy Carter appointed her to the board of the National Science Foundation (NSF), an independent government agency that makes policy recommendations to Congress and the president. President Ronald Reagan reappointed her to that post in 1986; from 1988 through 1991 she served as chairman of the NSF board. In 1991 President George Bush appointed her to the President's Council of Advisors on Science and Technology.

Good entered public service full-time in 1993, when President Bill Clinton appointed her undersecretary of commerce for technology—the highest-ranking technology post in the Department of Commerce. Good served as undersecretary for four years, during which time she stressed the need for government support of emerging technologies and worked extensively on the administration's Partnership for a New Generation Vehicle, also known as the "Clean Car" initiative. After the Republicans gained control of both houses of Congress in 1994, Good has said, she devoted much of her energy to fighting the efforts of GOP lawmakers to dismantle the Department of Commerce, which to some freshman legislators was a symbol of bloated government. She told Long in 1996 that she and her colleagues were spending "so much time defending what we are doing that it has been very hard to get progressive kinds of things done. We fought to defend the budget positions of these programs. We even fought to defend this office itself." Good explained that in her view, such attacks on the Commerce Department were misguided, because "the federal workforce looks like all the others [in the private sector]. There are a few leadership folks who work very hard, do very good things, are very imaginative and creative, who keep things going. There is a huge group, who—if they have good leadership and are inspired to do things—work very well, put out a solid day's work. And there is a small fragment who are not very good. I have seen exactly the same in my other two careers." In early 1996 Good served as acting secretary of commerce for nine days, after Ronald H. Brown, who had headed the department since 1993, was killed in a plane crash.

Good stepped down from her government post in 1997 to become Donaghey University Professor at the University of Arkansas at Little

Rock. At that time she was also named a managing member of Venture Capital Investors LLC. In early 2000 Good succeeded the renowned biologist and paleontologist Stephen J. Gould as president of the American Association for the Advancement of Science, the world's largest organization of scientists; she currently chairs the association's board of directors.

In her spare time, Good enjoys canoeing and studying Scottish history. After she accepted the job with UOP, her husband took early retirement and devoted himself to painting. In addition to their two sons, the two have four grandchildren.

Further Reading

Long, Janice R. "Mary Lower Good to Receive 1997 Priestley Medal." *Chemical and Engineering News*, May 13, 1996. Web. 13 August 2012. <http://pubs.acs.org/cen/hotarticles/cenear/960513/medal.html>.

Good, Mary L. Interview by Traynham, James G. *Mary L. Good.* Philadelphia, PA: Chemical Heritage Foundation, 2 June 1998. Print.

"Mary Lowe Good, PhD." *The Catalyst Series: Women in History.* Chemical Heritage Foundation. n.d. Web. 13 August 2012. <http://www.chemheritage.org/discover/>.

"Mary Good." *The Heinz Awards.* Heinzawards.net. 1 August 2004. Web. 13 August 2012. < http://www.heinzawards.net/recipients/mary-good>.

Gupta, Mahabir P.

Indian pharmacologist

Born: October 3, 1942; Gajsinghpur, India

Mahabir Gupta, a professor at the University of Panama, is one of the world's premier pharmacologists. He has been instrumental in organizing international efforts to research and catalog the medicinal properties of plants throughout Latin America, which contains many regions of great biodiversity. As the executive director of the Interciencia Association, a collective of mainly nongovernmental scientific groups from seventeen nations, Gupta tirelessly promotes scientific cooperation within the Americas and has been awarded the American Association for the Advancement of Science (AAAS) Award for International Scientific Cooperation as a result of his efforts.

Early Life and Education

Mahabir Prashad Gupta was born on October 3, 1942, in Gajsinghpur, India. He attended college at the University of Rajasthan, graduating in 1963. He then earned a master's degree at Banaras Hindu University, in Varanasi, an institution that teaches, according to its website, "the modern sciences and technology of the West," while promoting "the study of the Hindu Shastras [scriptures] and of Sanskrit literature as a means of preserving and popularizing the best thought and culture of the Hindus and all that was good and great." Gupta next traveled to the United States, where in 1971 he received a doctoral degree in pharmacology from Washington State University. The following year he was an Alexander von Humboldt research fellow at the University of Munich. The fellowship, named for the German researcher and explorer (1769–1859), was intended to allow promising young scientists to work on projects of their own choosing while living in Germany for six months to a year. Later in 1972 Gupta accepted a position as a professor of pharmacology at the University of Panama, with which he remains affiliated.

Life's Work

Gupta's interests lay largely in the field of ethnobotany, the study of how various cultures and regions use indigenous plants for purposes ranging from ritual to culinary to medicinal. Although ethnobotany draws anthropologists, historians, and sociologists, among others, because of increased interest over the last few decades in traditional and natural remedies, this area of study is of great concern to modern pharmacologists. One notable example of a plant-derived substance that was adopted by Western practitioners is quinine, extracted from the bark from the cinchona tree and used by residents of the Andes and Amazon highlands to treat fever, it was bought back to Europe by missionaries in the sixteenth century and proved invaluable in treating malaria. Another such example is curare, still used today by a few

> "Our scientific team is excited about the opportunity to share knowledge and collaborate with Unigen on a project designed to create novel and indigenous products addressing the health needs of the world."

Amazonian tribes to poison the tips of their arrows. In 1800 Alexander von Humboldt documented the complex method of preparing curare, whose active ingredient is derived from the plant species *Chondrodendron tomentosum*, and today synthetic versions are widely used in anesthesiology. Perhaps the best-known and most widely used medication derived from plant materials is aspirin: Willow-bark tea had been used by the early Greeks and Romans to treat fevers and pain, and Native Americans used a similar brew; it was not until the nineteenth century, however, that it was discovered that the ingredient responsible for the tea's healing powers was salacin, part of a family of compounds known as salicylates. A synthetic product, acetylsalicylic acid, was manufactured and marketed by Friedrich Bayer and Company and still sells well today.

At a meeting of the AAAS in 2000 Gupta revealed that he and his team had found that an extract from the bark of a tree in Argentina that showed promise as an anti-viral agent to fight HIV. They had formed an alliance with researchers from the University of Pamplona, in Spain, to develop the drug, but were not publicly naming the plant for fear of poachers. (At the same meeting a Swiss team announced that they had found an anti-fungal compound in the bark of the roots of an African tree. Interestingly, the bark of the above-ground portions of the tree showed no such medicinal properties.) Searching for such useful natural substances can be difficult, with hundreds of thousands of varieties of plants around the globe. Kurt Hostettmann, one of the Swiss researchers, told Ed Susman for *Biotechnology Newswatch* (March 6, 2000), "Some people do random screening, but with keen observation your rate of success in finding medicinal plants can be higher. You should take your time in selection of plants to test." Often, scientists will follow what Gupta calls "ethno-botanical leads" from tribal healers in Latin American countries, according to Dick Ahlstrom in the *Irish Times* (February 21, 2000). Such a tip led to his find in Argentina.

In addition to his teaching and research, Gupta is the executive director of the Interciencia Association, which includes members from Argentina, Bolivia, Brazil, Canada, Colombia, Costa Rica, Cuba, Chile, Ecuador, Jamaica, Mexico, Panama, Peru, Puerto Rico, Trinidad and Tobago, the United States, Uruguay, and Venezuela. For his work with this coalition—which besides fostering scientific cooperation between member nations also works towards the creation of societies for the advancement of science in countries where they do not yet exist—in 2003 Gupta was honored with the prestigious AAAS Award for International Scientific Cooperation, which came with a $5,000 prize.

In 2004, as the head of the University of Panama's School of Pharmacy and Center for Pharmacognostic Research on Panamanian Flora, Gupta entered into a partnership with Unigen Pharmaceuticals, an American company that researches thousands of varieties of plants in search of untapped medicinal uses. "Panama is among the top twenty-five most plant-rich countries in the world, with rain forests that provide a perfect level of light intensity, humidity, and temperature to yield unique and diverse plants containing the most biodiversity and high nutrient content," Qi Jia, Unigen's vice president of scientific affairs, told *Household and Personal Products Industry* (July 1, 2004). Gupta added, "Our scientific team is excited about the opportunity to share knowledge and collaborate with Unigen on a project designed to create novel and indigenous products addressing the health needs of the world." The collaboration is expected to significantly expand Unigen's ethnomedicinal plant library, which already included more than 3,000 plants and 5,000 plant extracts.

The term generally applied to the search for plant substances that can be turned into marketable drugs is bioprospecting. Because bioprospecting tends to be undertaken by scientists from developed countries who conduct their work in poorer, developing nations, it has been termed biopiracy by opponents who believe that the wealthier nations are stripping the developing world of its most useful resources—without redistributing its benefits to those countries. His proponents have pointed out that Gupta and his team have been scrupulous about following the United Nations' Convention on Biological Diversity. Signed in 1992 at a summit in Rio de Janeiro, the convention calls for the conservation of biological diversity, the sustainable use of its components, and the fair and equitable sharing of the benefits from the use of its resources.

Mahabir Gupta, who is an associate at the Smithsonian Tropical Research Institute in addition to his other appointments, has been married to Olga Victoria Barberena since June 29, 1974. They have three children: Rajesh, Sanjay, and Purni.

Further Reading

Gupta, Mahabir P., Handa S.S., and K. Vasisht."Biological screening of plant constituents, v ICS/UNIDO."*International Centre for Science and High Technology, Triestre* (2006): 114. Print.

Ioset, J.R., Maston, Andrew, Gupta, Mahabir, P., and Kurt Hostettmann. "Antifungal Xanthones from Roots of Marila Laxiflora." *Pharmaceutical Biology* 36.2 (1998): 103–106. Print.

"Mahabir P. Gupta." *AAAS Awards: 2003 Award Recipients*. AAS. n.d. Web. 13 August 2012. <http://www.aaas.org/aboutaaas/awards>.

"Tree Bark Could Fight AIDs." *BBC News*. 20 February 2000. Web. 13 August 2012. <http://news.bbc.co.uk/2/hi/sci/tech/specials/washington>.

"U.S. Meeting Acknowledges Tribal Healers." *Irish Times.com,* 2 February 2000. Web. 13 August 2012. <http://www.irishtimes.com/newspaper/ireland/2000>.

Harris, Eva

American geneticist

Born: August 6, 1965; New York, New York

The molecular biologist Eva Harris has likened her vision of a just society to the internal workings of a living cell. When one studies a cell, "if you understand how all the molecules work," it becomes clear that "there's this incredible energy conservation," Harris told Harry Kreisler for the Conversations with History interview series. In the same interview, as posted on the website of the Institute of International Studies at the University of California, Berkeley, she said, "There's a feedback loop that actually works. All the elements work together for the greater good of the whole. All of these really beautiful principles are played out, and that's how we are able to exist, all organisms. There are so many kinds of mottoes that we dream about in a just, human world, [which] are actually being played out every instant in our own bodies." Dubbed the "Robin Hood of biotechnology" by Claudia Dreifus in the *New York Times* (September 30, 2003), Harris has devoted herself to a practice that she calls "appropriate technology transfer": bringing cutting-edge technologies in molecular biology to developing countries, not as they were originally patented but in simplified and less expensive forms. Chief among her exports is polymerase-chain-reaction (PCR) technology, which, among other applications, can be used as a powerful diagnostic tool. In 1993 she founded and became director of the Applied Molecular Biology/Appropriate Technology Transfer (AMB/ATT) Program, which has brought life-saving technologies and knowledge to such countries as Nicaragua, Cuba, Guatemala, Ecuador, and Bolivia. According to the website of the Sustainable Sciences Institute, to date Harris and her coworkers have trained more than five hundred scientists in Central and South America in basic molecular-biology techniques for the diagnosis of infectious diseases.

In 1998 the AMB/ATT program became an arm of the Sustainable Sciences Institute, which Harris set up with money awarded to her as a

MacArthur Foundation fellow. She has also brought her humanistic approach to science to UC-Berkeley's School of Public Health, where she became an assistant professor of infectious diseases in 1998. "I see science as a social contract," Harris told Dreifus. "I do it to understand the biological functioning of cells and to understand disease to eradicate it. To me, science is about making a better world." At the conclusion of her interview with Kreisler, Harris said, "You have to be good [at what you do], but if you believe in something, follow your passion. I think that that's the most important thing. Don't let go of values, because that's what's going to matter. You're the one who's going to live with it for the rest of your life. If you can hold on to good values and make a positive impact on the world, then everybody else is going to benefit."

Early Life and Education

An only child, Eva Harris was born on August 6, 1965, in New York City. Her father, Zellig Sabbettai Harris, who immigrated with his family to the United States from the Ukraine at the age of four, was an internationally renowned linguist who founded the first linguistics department in the United States at the University of Pennsylvania in Philadelphia, his alma mater; students of his whom he profoundly influenced include the linguist Noam Chomsky. Originally a specialist in Semitic languages, Zellig Harris wrote more than a dozen books related to linguistics. Speaking of her father, Eva Harris told Harry Kreisler, "I spent a lot of time conversing with him about many, many things." Harris's mother, Naomi Sager, is a professor of computer science and linguistics at New York University.

Eva Harris spent her childhood shuttling between New York City and Paris, where her parents had many interesting friends. "One of the women who helped raise me fought in the Russian Revolution," Harris recalled to Kreisler. She noted that among her parents' other friends were people who had been members of the French Resistance, which fought Nazi Germany's occupation of France during World War II, and the German syndicalist movement, which advocated the union of all workers and governance of the nation by such a union. "I was very struck by their hearts of gold," Harris recalled. "People had given everything and risked their lives for other people. That made a big impression." In addition to being influenced politically by the liberal views of her parents and their friends, Harris, whose family did not own a television, began forming her political opinions at an early age through extensive reading. "When I was ten or so, I read all the classics—Orwell, and *All Quiet on the Western Front*. I always felt that even though I had a lot of politics around my life, I came to my own conclusions that, essentially, war is bad, and money's bad; you know, things like that. That was from my perspective as a six-year-old. I feel like I came to that on my own, even though, of course, it was within this context of [my parents and their friends]." She also told Kreisler, "My parents believed in me and let me do everything I wanted. I was an overachiever on my own, so they were always hoping I would fail at something once in a while." As a child she also spent much time engaged in crafts projects, constructing "tons of things," as she put it.

Harris attended Harvard University in Cambridge, Massachusetts, where, struggling to choose a major, she enrolled in many classes in the humanities—political economy, French literature, and art history, for example—before committing herself to biochemistry. During a summer project she met Jeff Schatz, a scientist from the University of Basel, whose efforts to bring advanced laboratory technologies to Guatemala impressed her. Harris received a bachelor of science degree from Harvard in 1987. She had been accepted to UC-Berkeley's graduate program in biology and had secured a National Science Foundation predoctoral fellowship when, in the same year, she decided to postpone graduate studies and instead travel to Nicaragua to teach biochemistry through a program that normally placed computer scientists. "They didn't really have any place to put me," Harris told Marcia Baringa for *Science* (November 25, 1994). "They literally just dropped me off at the Ministry of Health in Managua. There were roosters running everywhere. I barely spoke Spanish; nothing had prepared me for this. It was the most frightening experience of my life." Soon, however, Harris became fluent in Spanish (she had learned a variant of the language during an earlier visit to Spain). She taught a course in technical English and a seminar on molecular biology and helped to perfect a test for detecting endotoxins in blood plasma at a plasma factory. She also began working on a method for identifying the various strains of the *Leishmania* parasite, a protozoan that grows in the guts of sandflies and infects humans through sandfly bites. Depending on the particular strain (at least twenty-one species are known to affect humans), the resulting disease, leishmaniasis, has various manifestations, ranging from relatively benign skin lesions to acute systemic infections that may prove fatal, particularly in people whose health is already compromised because of malnutrition or other conditions. Each type of leishmaniasis must be treated differently.

After three months in Nicaragua, Harris returned to the United States to attend graduate school at UC-Berkeley. Working with Jeremy Thorner, she conducted research in yeast genetics; on her own, she continued to investigate Leishmania. The following summer Harris returned to Nicaragua and applied her newest research to the Leishmania problem. Once again, she was unable to find a reliable method for identifying different strains of the parasite. It was in connection to Leishmania that Harris got the idea for using PCR technology. Back in Berkeley, at her request, Christian Orrego, the director of the DNA lab at Berkeley's Museum of Comparative Zoology, taught Harris the PCR technique. With Orrego's help, Harris put together a five-day lab course in PCR and, after soliciting funding, began teaching it in Nicaragua in the summer of 1991. Working in very primitive labs, Harris succeeded both in reliably executing the PCR technique and in applying it to the identification of Leishmania strains. "I was on a high for months," Harris told Barinaga. "To see that this stuff was actually applicable and appropriate just blew my mind."

PCR involves isolating a string of DNA and placing it in a solution of DNA building blocks and other materials, where it can reproduce billions of times. In this way the DNA sequence is "amplified," and scientists can easily identify it, sequence it (that is, determine the exact order of the building blocks that make up that string), or clone it. Because the DNA of every species—including viruses, bacteria, and protozoans—is peculiar to that species, PCR is a relatively simple and accurate way of diagnosing illnesses caused by invasive organisms. According to ThinkCycle.org, "PCR can capture epidemiological information on strains and subtypes while simultaneously diagnosing a patient. . . . The speed of PCR and its utility in capturing finer data on the infectious agents make it an incredibly useful technique." After she returned to Berkeley, Harris continued work on her PhD thesis while investigating ways of using PCR to identify other disease organisms, such as the tuberculosis bacterium, several strains of diarrhea-causing *E. coli*, and a parasite implicated in malaria.

Life's Work

After she completed her doctorate, in 1993, Harris was accepted as a postdoctoral fellow to work under Stanley Falkow, a professor of microbiology, immunology, geographic medicine, and infectious diseases at Stanford University. However, she decided at the last minute not to enter the program. "I thought, 'Well, I can't just go on with my career and ignore all this excitement that we've engendered, so I'll take a year and make good on my promises, and then go back to my scientific life,'" Harris told Kreisler. Harris returned to Ecuador, where, with the help of her Latin American colleagues, she created the AMB/ATT program through the University of California, San Francisco. For several years the program kept afloat by means of an ill-funded core of volunteers. Harris's perennial worries about how she would scrape together enough money to continue its operations ended in 1997, when she won the $210,000 MacArthur grant. "It's like a fairy godmother appearing out of nowhere and saying, 'You can keep doing this work,'" Harris told Diana Walsh for the *San Francisco Examiner* (June 17, 1997). "It makes everything possible. I can continue to do this while I build up the research as well." With the grant money, Harris formed the Sustainable Sciences Institute of San Francisco to carry on her work in technology transfer. From 1997 to 1998 Harris continued her work in Latin America while also serving as an assistant adjunct professor at UC-San Francisco. In 1998, in collaboration with Nazreen Kadir, she published *A Low-Cost Approach to PCR: Appropriate Transfer of Biomolecular Techniques*, in which she outlined her simple and inexpensive methods for applying PCR technology to the field of public health. PCR and its associated supplies are often sold to laboratories in kits that make the process easier and more convenient but are also expensive and do not require an understanding of PCR's underlying concepts. Harris's knowledge-based technology transfer, as outlined in the book, involves teaching the fundamental principles of PCR so that practitioners need not rely on expensive kits to use the technology. For example, Harris explained to Dreifus that a silica substance often provided in kits to purify DNA samples normally sells for about $100 for a few milliliters: "The substance is essentially ceramic

> "I see science as a social contract. I do it to understand the biological functioning of cells and to understand disease to eradicate it. To me, science is about making a better world."

dust. So what we've done is buy a twenty-pound bag of ceramic dust for $5 at the hobby store. And you wash the stuff in nitric acid and sterilize it, and then you have thousands of tubes of that substance. We're not violating anything because the commercial manufacturers have their way of doing this, and we have ours." (Harris added that adopting patented technologies is not a violation of intellectual property rights if you do not charge a fee for your services; in the United States, however, many laboratory supplies are protected by patents, and it is illegal to use homemade versions of those products in hospitals.) Harris's handbook on transferring PCR technology is considered a useful and clear guide for practitioners in the public-health field. "What distinguishes this book from other manuals is its simple and practical approach to introducing the reader to the general principles and the theoretical basis of the technique, the practical details of the specific protocols, and the various ways to adapt PCR to a low-cost method," Olga Rickards wrote for *Human Biology* (August 2000). "Harris's worthy idea was not only to compile an essential work of reference for use as textbook and practical benchtop manual, but also to provide a model for the transfer of other technologies to laboratories working in vastly different situations."

Harris joined the faculty of UC-Berkeley in 1998. She is a member of the university's Graduate Group in Infectious Diseases and Immunity. She also serves as a codirector of the Fogarty International

Emerging Infectious Diseases Training Program, an activity of the National Institutes of Health. In the 2003–2004 academic year, she taught graduate-level courses in infectious-disease research in developing countries and in molecular parasitology. She has published articles in such professional publications as the *Journal of Virology*, the *Journal of Parasitology*, the *American Journal of Tropical Medicine and Hygiene*, the *Journal of Clinical Microbiology*, *Clinical and Diagnostic Laboratory Immunology*, and *Emerging Infectious Diseases*, a journal of the Center for Disease Control and Prevention, an arm of the US Department of Health and Human Services. Through the Sustainable Sciences Institute, she has continued to bring PCR and other technologies to countries throughout Latin America. Currently, she is conducting research on, among other subjects, the dengue virus, which is transmitted by mosquitoes and infects roughly 100 million people annually. Harris herself contracted dengue fever in 1995 in Nicaragua; according to the website of UC-Berkeley's Health Sciences Initiative, she diagnosed the disease in its early stages. (Unlike Harris, the vast majority of victims do not have access to treatment.) She decided to study the virus because of its pervasiveness. She recalled to Kreisler, "Everyone said, 'Eva, work on this. You always said you were going to follow the urgency.' I said, 'I'm not a virologist'; then I said, 'Well, I guess, I'm going to become one.'" The "basic science" aspect of her work involves, as she explained to Kreisler, "testing certain antiviral compounds to see if they are effective against dengue virus. If we understand how the virus replicates, we can identify pieces of the virus which are important to knock out, to create an attenuated strain for a vaccine."

Eva Harris is a member of the American Association for the Advancement of Science (AAAS) and the American Society for Microbiology (ASM). In 2001 she won an investigator's award from the Pew Scholars Program in the Biomedical Sciences. The next year she won the Prytanean Faculty Award, which honors an outstanding UC-Berkeley female assistant professor. Also that year the Ellison Medical Foundation pledged up to $750,000 to support Harris's attempts to develop a so-called mouse model for study of the dengue virus. In 2003 she and her UC-Berkeley colleague P. Robert Beatty won a Doris Duke Innovation in Clinical Research Award to support their development of an "ImmunoSensor" for HIV infections. Harris is currently the faculty director of the Center for Global Public Health at UC-Berkeley.

Further Reading

Dreifus, Claudia. "A Conversation With Eva Harris; How the Simple Side of High-Tech Makes the Developing World Better." *New York Times* 30 Sept. 2003: F3. Print.

Kreisler, Harry. "Conversation with Eva Harris." *Conversations with History.* Regents of the University of California, 15 Mar. 2001. Web. 9 Aug. 2012.

Marris, Emma. "Straight Talk From...Eva Harris." *Nature Medicine.* Nature Publishing Group, 2007. Web. 9 Aug. 2012.

Heath, James R.

American chemist

Born: 1961, England

"Our ultimate goal is to build a computer with approximately the power of 100 high-end workstations on a platform the size of a grain of sand," the chemist James R. Heath wrote for the website of the Chemistry Department at the University of California, Los Angeles (UCLA), where he worked from 1994 to 2002. Currently the Elizabeth W. Gilloon Professor of Chemistry at the California Institute of Technology (Caltech), Heath is a leader in a groundbreaking field known variously as nanoscience, nanotechnology, and molecular nanotechnology; the specialty is also known as molecular electronics, moletronics, and molecular engineering. Heath and other practitioners in the field focus on the formation, study, and manipulation of molecules or bits of matter only a few molecules in width. Among Heath's projects is the chemical synthesis of a computer, one whose dimensions would be described in terms of molecular measurements (Angstrom units or microns, one Angstrom unit equaling 0.0001 microns or one ten-billionth of a meter). Such computers would perform the same operations as even the most sophisticated computers now in use, but they would not rely on the etched circuits of silicon chips, as do present-day computers. "Today, you take a material [silicon] and you whittle away almost all of it using lithography," Heath explained to Gary Taubes for *UCLA Magazine* (Spring 2000). "It's like sculpting. You come with a chisel and you take away all the stuff you don't want. I started thinking about how to build a computer by bringing small amounts of stuff together, molecules at a time. And I realized there's no reason why it can't work." A molecular computer might be far cheaper to construct and far more efficient and inexpensive to operate, compute billions of times faster, and have vastly more memory than contemporary computers, and it could be used in ways that have until now only been dreamed of. To cite one example, such a minuscule device might serve as a sensor that, while traveling in a person's bloodstream, could identify foreign bacteria and alert its host to potential trouble. "What once seemed like science fiction is now looking more and more like science!" Heath exclaimed on his UCLA webpage. He told Gary Taubes, "I really want to make this computer happen. I want to see it get built. It would be great if it changed the world, and I think we have a legitimate chance."

In 2000 Heath helped to set up the California NanoSystems Institute (CNSI), a joint endeavor of UCLA, UC-Santa Barbara (UCSB), private industry, and the state of California. The institute specializes in nanoscience and nanotechnology, the prefix "nano" meaning "one-billionth" (thus, a nanometer is one-billionth of a meter—as small, proportionately, as a quarter-inch is to the distance from Los Angeles to New York). The institute's goal, as Jayne Fried put it in *Small Times* (March 22, 2002), is to "build and examine natural systems atom-by-atom and molecule-by-molecule." Since he joined the Caltech faulty, in 2002, Heath has continued to collaborate on CSNI projects, one of which has led to the recent invention of a way to make ultra-high-density nanowire lattices and circuits. The lattices consist of grids of crossed nanowires, in which each cross, or junction, "represents the elements of a simple circuit!" as Jacquelyn Savani wrote for a UCSB press release (March 13, 2003). One square centimeter of such a grid has enough room for more than 1011 junctions. At Caltech, Heath leads a research group that includes six postdoctoral researchers, a dozen graduate students, and a few undergraduate students. Their research projects involve nanosystems biology, molecular electronics, and fundamental nanoscale physics and chemistry dealing with such entities as quantum dots, or semiconductor nanocrystals. In 2003, along with six other distinguished scientists from Caltech, UCLA, and the Seattle-based Institute for Systems Biology, Heath formed the NanoSystems Biology Alliance, whose mission is to develop nanotechnological tools, including tools used in the new science of microfluidics (the study of minuscule amounts of fluids) for use in diagnosing disease and in "attacking challenges in cancer and immunology," according to the alliance's website. Before he reached his fortieth birthday, Heath received several prestigious awards for his achievements, among them the Feynman Prize in Nanotechnology, in 2000. His honors also include, in 2002, the *Scientific American* 50 Award, which recognized him among "visionaries . . . whose recent accomplishments point toward a brighter technological future for everyone."

Early Life and Education

James R. Heath was born in 1961 in England to American parents; his family moved to the United States shortly thereafter. He was raised near Houston, Texas. As a child he passionately loved music; he studied trumpet in grade school and later learned to play the guitar, saxophone, and violin. During his undergraduate years, at Baylor University, in Waco, Texas, he entertained at clubs; he did so while in graduate school as well. After he received a bachelor's of science degree in chemistry in 1984, he entered Rice University in Houston, Texas, where he pursued a doctoral degree under the guidance of the chemist Richard E. Smalley. On his first day in Smalley's lab, he operated a huge machine, designed by Smalley, that used lasers to vaporize and subsequently analyze a class of unusual molecules known as clusters; he became so fascinated that he remained in the lab until 3:00 a.m., when he awakened Smalley with a phone call, asking how to turn off the machine. "I knew I had found the right place to be," he told Gary Taubes. At that time Smalley was collaborating with another Rice University chemist, Robert F. Curl. A year later the group expanded to include a collaboration with the British chemist Harold Kroto, of the University of Sussex in England. Their research into the nature of interstellar matter led them—with the help of Heath and two other graduate students—to the discovery of a new spherical molecule composed of sixty atoms of carbon, whose structure resembles that of a soccer ball. The stability of this new form of carbon was reminiscent of the architectural stability of the geodesic dome, invented by the American engineer and architect Buckminster Fuller. The compound, named buckminsterfullerene and nicknamed the buckyball, proved to be one of a family of all-carbon compounds, dubbed the fullerenes. Smalley, Curl, and Kroto won the Nobel Prize in 1996 for their discovery. For his contributions to their work, Heath earned the Rice University Graduate Award in 1987. He completed his PhD in 1988.

Life's Work

Soon afterward, as a Miller Postdoctoral Fellow, Heath came to the University of California, Berkeley. There, he worked in the laboratory of the eminent chemist Richard J. Saykally, whose specialties include laser spectroscopy, a technique for detecting the presence and structure of molecules. To Heath, as quoted by Taubes, Saykally was "prob-

> **"What once seemed like science fiction is now looking more and more like science!"**

ably the world's best spectroscopist," while Saykally told Taubes that he considered Heath "the most brilliant experimentalist" with whom he had ever collaborated. Heath's extraordinary skills notwithstanding, his attempts to secure a faculty position for the following year met with failure at the twelve colleges at which he interviewed, because his research proposals were "a little too futuristic for most chemistry departments," as Saykally explained to Taubes. Disheartened, Heath told Saykally that he would abandon chemistry altogether rather than accept a job as a traditional industrial chemist. Thanks to Saykally's powers of persuasion, Heath remained at Berkeley as a postgraduate researcher.

When he next applied for work elsewhere, Heath refrained from divulging his true research ambitions, emphasizing instead a much more conservative research program. That approach succeeded: in 1991 he landed a job at IBM's T. J. Watson Research Labs in Yorktown Heights, New York. A series of financial problems at IBM in the early 1990s delayed the purchase of lasers and other equipment in Heath's lab at IBM, and his frustration led Heath to consider changing his scientific goals. "But IBM emphasizes big, important problems in industry and Jim learned that the big, important problem of the time was how to make silicon and germanium nanostructures and nanowires," Saykally recalled to Taubes. "And Jim just got out a bunch of beakers, read the literature, and figured out how to solve the big problem of the day, all the time while waiting for his physical chemistry lab to come through. And he did it ahead of all the top people in the field. Just like that. He never did uncrate his lasers."

In 1994 Heath left IBM to join the Department of Chemistry at UCLA. In his UCLA labs he began working with tiny particles of silver, called silver quantum dots. Using them as "artificial atoms," he built from them a unique set of solids, appropriately called "artificial solids." Those solids turned out to be remarkable systems for investigating what are known as quantum-phase transitions. A phase transition is, for example, water freezing from a liquid into solid ice. A quantum phase transition relates to a similarly dramatic change of properties of the electrons of a material.

In 1996 R. Stanley Williams, a one-time UCLA chemistry professor who had left UCLA to join Hewlett-Packard, enlisted Heath to help him set up a basic research department in physical sciences at Hewlett-Packard Laboratories in Palo Alto, California. (Williams directs the department, now known as the Quantum Structures Research Initiative.) At Hewlett-Packard, Heath and Williams met Philip J. Kuekes, who had worked for the company for three decades as an architect, or designer, of computers, and the three scientists began discussing the possibility of building a molecular-based computer. Although they were not quite sure what such a computer would look like, they understood that anything that included molecules would likely also include defective components. Before Heath's arrival, Kuekes had built an experimental computer, called Teramac, that contained approximately 220,000 hardware defects, any one of which, if present in a conventional computer, might have caused major problems with its functioning. Teramac, able to circumvent those defects, was a supercomputer, far outperforming most other computers of the day. Heath, Kuekes, and Williams described Teramac's remarkable abilities in *Science* (June 12, 1998), in a paper in which they hypothesized that such tolerance of defects could also be incorporated in chemically synthesized nanoscale electronic components.

Developing Molecular Electronics

Back at UCLA, Heath began experimenting with a family of molecules called rotaxanes and catenanes, which had been developed by his UCLA colleague J. Fraser Stoddart. These unique compounds consist of two molecules that are linked mechanically but can behave independently. Catenanes appear like two interlocked Olympic rings, while the rotaxane molecule looks something like "a barbell with a ring around it," as Ashley Dunn put it in the *Los Angeles Times* (July 16, 1999), with the "barbell" part being a linear molecule (made up of a naphthyl group on one side and a tetrathiafulvalene, or TTF, group on the other) and the ring being another molecule, called a tetracationic cyclophane; bulblike structures at each end of the "barbell" prevent the ring from sliding off. Particular external stimuli can cause the ring to move from the naphthyl side of the barbell to the TTF side; in that sense, the rotaxane molecule is "switchable." Heath realized that the "switchability" of rotaxanes theoretically made possible the use of such molecules in a chemically assembled computer, where the rotaxane molecules' behavior would replicate the operations of digital switches (the on-off mechanisms that form the foundation of modern computers). Digital switches are created by engraving silicon wafers with beams of light by means of photolithography; for that reason, such circuits can be no smaller than the wavelength of light. Many molecules, however, including those of the rotaxanes, are far smaller than the wavelength of light; therefore, chemical switches made with rotaxanes could be far smaller than digital ones. Heath and Stoddart formed a team to come up with ways to merge Kuekes's computer designs with the rotaxane switches.

On July 16, 1999, in a paper in *Science* entitled "Electronically Configurable Molecular-Based Logic Gates," members of Heath's team announced the creation of basic computing circuits using molecular components. They used rotaxane molecules that acted as switches that could be switched from open to closed only once, but this allowed them to demonstrate the validity of their ideas of how the defect-tolerant Teramac machine could be applied to molecular electronics. A year later, in an article in *Science* (August 18, 2000), Heath and Stoddart reported on the first molecular switch, based on a catenane that could be cycled between the open and closed states repeatedly, similar to more traditional silicon-based transistor switches, and they subsequently reported on a series of cyclable catenane and rotaxane switches. For these switches, the molecules were sandwiched between two electrodes. For the rotaxane case, when voltage was applied to the molecular device, the cyclophane ring moved from surrounding the TTF site to surrounding the naphthyl site. Further investigation showed that the electrical resistance of the naphthyl group was ten to twenty times lower than that of the TTF group; thus, when the ring was over the naphthyl group, the "switch" was considered "on"; the ring's move to the side with the TTF group turned the switch "off." As Kenneth Chang wrote for the *New York Times* (July 17, 2001), "groups of those molecules could be switched back and forth between the on and off positions." In another article in *Science* (March 13, 2003), Heath and Pierre Petroff, from the UCSB Department of Electrical and Computer Engineering, and his collaborators announced their invention of a method to produce ultra-high-density nanowires and nanowire circuits and, with it, a way to make memory chips on a grid, or array, of wires arranged in two planes, one perpendicular to the other, in a so-called cross-bar architecture; the molecular switches lie at the intersections of the wires. That breakthrough brought the molecular computer yet another step closer to reality. In a summary of their accomplishments that appeared on a press release (April 30, 2002) from Nanosys Inc., which acquired the exclusive rights to "intellectual property related to cross-bar array architecture" from UCLA in 2002, Heath said, "By linking molecular switches and wires, we created highly effective logic gates with true molecular dimensions. . . . In the lab, we have already made working 16-bit memory circuits. Within a few years, we will have in place working prototypes for a wide array of nano-based devices."

The Nanosys press release appeared one month after Heath announced his intention to leave UCLA to work at Caltech in Pasadena. He had become exhausted with trying to run, concurrently, the California NanoSystems Institute and a large research group. "At Caltech, I just do research and teach and, simply put, Caltech is a better place for doing research," he told *Current Biography*. Heath is a founding member of Nanosys's scientific advisory board. In December 21, 2001, *Science* magazine labeled the nanowire and crossbar-memory circuit work of Heath and Charles M. Lieber, a Harvard chemistry professor and one of the scientific founders of Nanosys, as the scientific breakthrough of the year 2001.

In addition to the Feynman Prize and the *Scientific American* 50 Award, Heath has won a National Science Foundation Young Investigator Award (1994), the Seaborg Award (1997), the Jules Springer Prize for Applied Physics (2000), and the Raymond and Beverly Sackler Prize in the Physical Sciences (2001). He earned Packard and Sloan fellowships in 1994 and 1997, respectively, and was elected a fellow of the American Physical Society in 1999. In 2009, Health was named one of the seven most powerful innovators by *Forbes* magazine. Heath often arrives at his laboratory at 4:00 a.m. and puts in fourteen- to sixteen-hour workdays. On his Caltech website, he is shown playing the guitar in his lab. Heath and his wife have two children.

Further Reading

Brown, Eryn. "Building Chips. One Molecule at a Time." *Fortune* 14 May 2001: 166. Print.

Dunn, Ashley. "Molecular Computer Takes a Major Stride." *Los Angeles Times.* Los Angeles Times, 16 July 1999. Web. 9 Aug. 2012.

Taubes, Gary. "Speed Demon." *UCLA Magazine*. The Regents of the University of California, Spring 2000. Web. 9 Aug. 2012.

Hershko, Avram

Israeli biochemist

Born: December 31, 1937; Karcag, Hungary

On October 6, 2004, the Royal Swedish Academy of Sciences awarded Avram Hershko, Aaron Ciechanover, and Irwin Rose the Nobel Prize in Chemistry "for the discovery of ubiquitin-mediated protein degradation," the Nobel Foundation website stated. "Thanks to the work of the three Laureates it is now possible to understand at molecular level how the cell controls a number of central processes by breaking down certain proteins and not others." Hershko was not at home to receive the initial call from Sweden, having taken his four granddaughters on a picnic to celebrate the Jewish harvest festival of Sukkot. He received the news later in the day when a cousin, who heard the announcement on the radio, called him on his cell phone. In an interview posted on the Nobel Foundation website, Hershko declared, "I am very happy. Very happy for my family, for my institution, my country, and for myself also." He was particularly pleased that Ciechanover and Rose were able to share the award with him. After Hershko explained to his four granddaughters what the prize was, the girls wondered what they would wear to the award ceremony in Sweden. "I told them, 'You have to wear a white dress because you meet the king and queen.' They were very impressed by this," Hershko recalled to David Brown. Hershko used the exposure generated by his winning the Nobel Prize to urge the Israeli government to offer more support for education.

Early Life and Education

Avram Hershko was born Ferenc Hershko on December 31, 1937, in Karcag, Hungary, a town about eighty-five miles southeast of Budapest. As Jews, Hershko and his family were greatly affected by the Nazi occupation of their country during World War II. His father, Moshe Hershko, was sent to the Russian front as a slave laborer, captured by the Russian army, and used as a forced laborer in the Soviet

Union until 1946. (For four years, no one in Hershko's family knew if he was alive.) Hershko, his mother, Shoshana, and brother, Chaim, were initially confined to a local ghetto, which was for most Jews a stopover on the way to Auschwitz. Instead, the Hershkos were eventually moved to a labor camp in Austria, where they managed to survive the war, along with three aunts and his paternal grandparents. (His maternal grandparents were killed in the Holocaust.) After the war, the family was reunited when Hershko's father returned from the east.

In 1950, after living in Budapest for three years, the Hershkos immigrated to Israel, settling in Jerusalem, where Hershko's father earned a living as a teacher; he later wrote mathematics textbooks. Hershko studied at the Ma'aleh School in Jerusalem, and subsequently attended the Hadassah School of Medicine, part of Hebrew University, receiving his MD in 1965. For the next two years, Hershko served in the Israeli Defense Forces as a physician. Subsequently, he returned to the Hadassah School of Medicine and obtained his PhD in biochemistry in 1969. Early on in his studies, Hershko resolved to pursue a career in academia and research rather than in clinical practice. "Research interested me more, and I thought clinical medicine would be routine," he explained to Judy Siegel-Itzkovich for the *Jerusalem Post* (October 8, 2004). From 1969 to 1971 Hershko served as a postdoctoral fellow with Gordon Tomkins, a renowned biochemist and expert in the study of hormones, at the University of California, San Francisco; it was then that he developed an interest in protein degradation. After returning to Israel, Hershko took a post as an associate professor at the Technion-Israel Institute of Technology in Haifa; he became a full professor there in 1980 and a Distinguished Professor in 1998.

Life's Work

At the Technion-Israel Institute, Hershko conducted the research into the process of protein degradation—how a cell breaks down unwanted proteins—that ultimately led to the Nobel Prize. He was assisted in this work most notably by two colleagues with whom he would later share the prize—Rose, whom Hershko first met at a conference of the National Institutes of Health in 1976, and Ciechanover, who that same year came under Hershko's tutelage as a doctoral student at the

Technion-Israel Institute. In the ensuing years Hershko and Ciechanover often took sabbaticals in order to work with Rose at the Fox Chase Cancer Research Center in Philadelphia, Pennsylvania, where the trio conducted some of its groundbreaking research. In opting to examine protein degradation, Hershko and his colleagues were delving into fairly obscure territory: researchers up until that point had focused on how a cell creates proteins, not how those proteins are broken down and destroyed. "Nobody else seemed interested in this then, but I thought it was important," Hershko recalled to Siegel-Itzkovich for the *Jerusalem Post* (May 20, 2001).

The chief question that Hershko and his associates tried to answer was why the process of protein degradation within a cell required energy, in the form of adenosine triphosphate (ATP), while other forms of protein degradation had no such energy requirement. Previous work in the laboratory of Alfred Goldberg, a respected biochemist and professor at Harvard Medical School, had shown that using extracts from immature red blood cells, researchers could reproduce ATP-dependent protein degradation in a test tube. Hershko and Ciechanover found that such extracts could be divided into two parts, each of which was dormant by itself, but when combined, initiated ATP-fueled protein degradation. The active ingredient in one of these parts, they discovered, was a protein they termed ATP-dependent proteolysis factor 1 (APF-1). Soon afterward the research team of Keith Wilkinson, Arthur Haas, and Michael Urban discovered that APF-1 was, in fact, ubiquitin, a molecule first isolated in 1975 and found to be present in every organism more complex than bacteria. Up until these revelations, researchers did not understand what purpose ubiquitin served.

In two seminal papers published in 1980, Hershko, Ciechanover, Rose, and Hannah Heller, a graduate student in Hershko's laboratory at the Technion-Israel Institute, described how APF-1 formed stable chemical bonds with a number of different proteins found in immature red blood cells; in the second paper the researchers proved that "APF-1 molecules could be bound to the same target protein, in a process known as polyubiquitination," as the process is explained on the Nobel Foundation website (NobelPrize.org). When bound to the target protein, the APF-1 molecule serves as a kind of "kiss of death," effectively marking that particular protein for destruction. "At a stroke, these entirely unanticipated discoveries changed the conditions for future work," an article on the Nobel Foundation website explained. "It now became possible to concentrate on identifying the

> **"I didn't go into science and medical research to win prizes or make money."**

enzyme system that binds ubiquitin to its target proteins." Given ubiquitin's presence in so many different organisms, the scientists realized that the substance must play a vital role in the overall function of a cell. Moreover, Hershko, Ciechanover, and Rose theorized that since ATP was required for polyubiquitination, the amount of ATP involved would serve as a sort of calibrator through which the cell could "control the specificity of the process," as the Nobel Foundation website stated.

The Multistep Ubiquitin-Tagging Hypothesis

Between 1981 and 1983 Hershko, Ciechanover, and Rose developed "the multistep ubiquitin-tagging hypothesis," which described the sequence of reactions by which a particular protein is destroyed. The process was catalyzed by three enzyme activities characterized as E1, E2, and E3. In the first step, the E1 enzyme, spurred by ATP, activates the ubiquitin molecule. From there the ubiquitin molecule is forwarded to the E2 enzyme to which it then bonds. In the third step, according to the Nobel Foundation website, "The E3 enzyme can recognize the protein target which is to be destroyed. The E2-ubiquitin complex binds so near to the protein target that the actual ubiquitin label can be transferred from E2 to the target." In the fourth step the ubiquitin-marked protein is released by the E3 enzyme, a process that is replicated until a chain of ubiquitin molecules are connected to the target protein. This chain, it was later learned, is then identified by the proteasome—a barrel-shaped structure within a cell that is responsible for breaking up or degrading proteins—and disconnected from the protein target, which is then consumed by the proteasome. (The proteasome had not yet been discovered when Hershko and company came up with their hypothesis; the scholarship of Sherwin Wilk and Marian Orlowski revealed the existence of the proteasome, the function of which was more fully explained by Martin Rechsteiner and Alfred L. Goldberg.) While the research of Hershko and his associates has since been validated, their initial findings were greeted with skepticism. "For quite a while, colleagues dismissed, even ridiculed, our ideas," Hershko recalled to Siegel-Itzkovich. "They said it wasn't logical for one protein to connect to another protein." Through another innovative experiment, Hershko and his colleagues further honed their understanding of the ubiquitin system. While studies exploring polyubiquitination had, until that point, been conducted in "cell-free systems," as noted on the Nobel Foundation website, "To be able to study the physiological function of ubiquitin-mediated protein degradation as well, [Hershko] and his coworkers developed an immunochemical method." The researchers employed ubiquitin antibodies to isolate ubiquitin-protein-conjugate "from cells where the cell proteins had been pulse-labeled with a radioactive amino acid." The data produced from this experiment proved the existence of the ubiquitin system in living cells.

The medical implications of this new understanding of protein degradation have been considerable: certain forms of cancer, as well as Parkinson's Disease and cystic fibrosis, are caused by flaws in the enzymes responsible for bonding ubiquitin to or detaching it from its protein target. "If proteins are not degraded at the right time," Hershko remarked to Siegel-Itzkovich for the *Jerusalem Post* (October 30, 2003), "the cell continues to divide unchecked. This is what happens in many cancer cells; something has gone wrong in the ubiquitin system so there is no control over cell division." Medicines that make use of Hershko's scholarship in that they influence the destruction of certain proteins are just beginning to come into circulation. For instance, Velcade—an injectable drug that combats multiple myeloma, a bone marrow cancer, by interfering with the functioning of the proteasome—has recently been approved by the US Food and Drug Administration (FDA); many more medicines of a similar nature are expected to undergo clinical trials in the near future. "[Velcade] is the first drug specifically targeted against the ubiquitin system," Hershko told Siegel-Itzkovich. "I am sure that many other new drugs will be discovered which are targeted against specific processes that go wrong in the ubiquitin system in different types of cancer. These include cancer of the colon, breast, prostate, and melanomas." Despite the emerging impact of his scholarship on the field of medicine, Hershko does not hold any patents related to his discoveries—patents that conceivably could have made him quite wealthy. "I didn't go into science and medical research to win prizes or make money," he affirmed to Siegel-Itzkovich.

Hershko has continued to explore the inner workings of the ubiquitin system and its role in protein degradation since his initial discoveries in the 1970s and 1980s. For over a decade he has spent his summers at the Marine Biological Laboratory in Woods Hole, Massachusetts, where he has studied ubiquitin's function in the cell-division cycle, using—in lieu of immature red blood cells—the eggs of a certain type of clam called spisula. His work has focused specifically on the ubiquitin-ligating complex of the clam eggs as well as of cultured human cells. This study of the "mechanisms of cell division . . . may provide novel insights into their aberration in human cancer," Hershko told a writer for *AScribe Newswire* (October 6, 2004).

In addition to the Nobel Prize, Hershko has earned a number of other awards during his career. Among his more noteworthy honors are Israel's 1987 Weizmann Prize for Science; the 1994 Israel Prize in Biochemistry; the Gairdner Foundation of Canada's 1999 International Award; the 2000 Alfred P. Sloan Jr. Prize; and the 2000 Albert Lasker Award for Basic Medical Research, an honor often bestowed upon future Nobel laureates. In 2001 Hershko was the recipient of both the Wolf Prize in Medicine and Columbia University's Horwitz Prize. Hershko has been a member of the Council of the European Molecular Biology Organization since 1993 and the Israel Academy of Sciences since 2000; in 2003 he was appointed a foreign associate of the US National Academy of Sciences. Hershko currently serves as a Distinguished Professor of biochemistry on the Rappaport Faculty of Medicine at the Technion-Israel Institute of Technology.

Personal Life

Hershko's wife, Judith, is a Swiss-born biologist who often works with her husband in the laboratory. They have three sons—Danny, a surgeon, born 1964; Yair, a computer engineer, born 1968; and Oded, a medical student, born 1975—as well as six grandchildren. Speaking of his plans for the future in the interview posted to the Nobel Foundation website, Hershko remarked, "I try to do an experiment every day. . . . And, I would like to continue with that because it's really exciting. . . . I think I can still contribute." He continued, "Not in the same big way as twenty-five years ago, but still contributing and then still having a lot of fun at the [laboratory]."

Further Reading

Brown, David. "Israeli, American Chemists Win Nobel." *Washington Post* 7 Oct. 2004: A3. Print.

Dreifus, Claudia. "The Body's Protein Cleaning Machine." *New York Times.* 19 June 2012: D2. Print.

"Hungarian-born Nobel Prize Winner to visit Hungary." Europe Intelligence Wire. 8 Oct. 2004. Print.

Jones, Monty

Sierra Leonean plant geneticist and rice breeder

Born: Freetown, Sierra Leone

Rice is a staple part of the diet in West and Central Africa, and one species, *Oryza glaberrima*, has been grown there for thousands of years. It has adapted to the harsh conditions of the region—resistant to a plethora of indigenous weeds and pests and able to survive periods of drought. *O. glaberrima*, however, is an incredibly low-yielding species, so Africa's farmers often turn instead to *Oryza sativa*, an Asian species of rice, which has much higher yields but is not resistant to the local weeds and pests. A cycle that is detrimental to both the environment and the economy is thus begun: because farmers cannot afford to use the commercial pesticides and fertilizers that high-yielding Asian rice requires for optimum growth, they clear new fields every few years using slash-and-burn methods. Because land is scarce and the farmers poor, fields that should stay fallow for at least a decade are recleared and replanted before fully recovering. Neither option—low-yielding African rice or high-yielding but costly Asian rice—is ideal, and rice must frequently be imported to the area from other countries, at an estimated cost of $1 billion annually. Scientists have long reasoned that combining strains of Asian and African rice would result in hardy, high-yielding crops but were continually thwarted in their attempts to do so. Finally, in 1994, a team of West African scientists, led by Monty Jones, succeeded in developing a species known as Nerica (the name is sometimes rendered NERICA and stands for New Rice for Africa)—which is not only hardy, quick growing, and high yielding, but also contains more protein than either of its parent strains. "It is a miracle crop," Kanayo Nwanze, the head of the West Africa Rice Development Association (WARDA), told Ernest Harsch for *Africa Renewal* (January 2004). "It will mean more food on each household's table and more money in the farmers' pockets. This will in turn contribute to food security and poverty reduction." For his efforts, Jones was given the 2004 World Food Prize, which in the world of agriculture holds similar prestige to a Nobel Prize.

Despite the potential benefits of Nerica, not all West Africans have been able to take advantage of the new species. Many farmers lack the credit needed to purchase seeds and fertilizer, and poor roads and farming infrastructure make it difficult for them to transport their crops to the market. In addition, inadequate storage facilities fail to protect yields from floods and droughts, and Nerica seed that mixes with regular rice seed often becomes less productive. As of October 2007, the Africa Rice Center estimated that a mere 200,000 farmers were growing Nerica on 5 percent of the land capable of supporting the crop. "If we don't develop the infrastructure, there's no way we'll attain the Green Revolution," Jones told Celia W. Dugger for the *New York Times* (October 10, 2007). "How do you bring the Nericas to farmers? How do you get farmers to know the seeds exist?" In an effort to bring the new varieties of rice to a greater number of people, the nonprofit Alliance for a Green Revolution, for which Jones serves as a board member, has said it will invest $23 million in the distribution of seeds.

Early Life and Education

Monty Patrick Jones was born in about 1951 in Freetown, the capital of the African nation of Sierra Leone. The product of a strict, Catholic upbringing, he briefly considered entering the priesthood. He dreamt, however, of bettering the world in more practical ways: Although his parents were both white-collar professionals and he had experienced little contact with farmers, he realized that the field of agriculture might provide both a viable career and a chance to do good. In 1974 he earned a bachelor's degree in agricultural science from Njala University.

Upon his graduation he began working with the WARDA, also known as the African Rice Center, an intergovernmental research association and the foremost authority on rice species in sub-Saharan Africa. He was appointed deputy director of the organization's Mangrove Swamp Rice Research Project, based in Rokupr, Sierra Leone. He was particularly fascinated by the saline conditions in the swamps, which enjoyed only a three-month salt-free period per year, because of a near-constant incursion of sea water. "The only varieties [of rice] which were sufficiently tolerant of the salt to be able to grow there were glaberrimas," he is quoted as saying in the WARDA's 1999 annual report. He also noted the rice's ability to thrive in marginal, upland environments. "Thus, it seemed that the glaberrimas had genes

> "When Africa breaks free from the grip of poverty and famine—as it now looks poised to do—Jones . . . will have played a pivotal role."

for resistance or tolerance to local stresses, such as soil acidity, iron toxicity, and blast disease, which were not available among the sativas," he said.

In the late 1970s the United Nations Food and Agriculture Organization awarded Jones a research fellowship that enabled him to attend Birmingham University in the United Kingdom. He received a master's degree in plant genetics in 1979, and remained at the school, from which he earned a doctoral degree in plant biology in 1983. Jones then returned to his native Sierra Leone, with the goal of combating Africa's food shortages. (His resolve was strengthened by the so-called "rice riots"—episodes of urban unrest and violence—that were plaguing portions of the continent during this period.)

Life's Work

From 1987 to 1990 Jones worked with the International Institute of Tropical Agriculture (IITA), a Nigerian-based research organization specializing in food production in the humid and sub-humid zones of Africa. In 1991 he was named head of the upland rice-breeder and rain-fed rice program at the WARDA's main research center and headquarters in Cote d'Ivoire. There he cultivated samples of *O. glaberrima* collected from gene banks at the IITA and the International Rice Research Institute (IRRI), a facility based in Asia. From 1,500 samples, Jones and his research team selected and tested forty-eight particularly high-quality varieties of the African rice as potential candidates for crossbreeding with *O. sativa*. Only eight were deemed potentially compatible for crossbreeding.

The WARDA's 1991 report explains: "Previous attempts to cross the two species had met with failure—the offspring were infertile, or else the few fertile seeds were not noticed among the infertile ones. Jones's determination to succeed led the team to unprecedented lengths: even the eight compatible varieties set very little seed in the first generation of the cross." The report quotes Jones as saying, "We were collecting five seeds or less from apparently sterile plants, but what a goldmine we'd uncovered." The seeds were carefully cultivated and the plants grown from them were then "backcrossed" with the *O. sativa* parent. The process was then repeated. While the fertility rates of the plants increased, Jones was still unable to produce what are termed "true-breeding" parental lines. In 1993 Jones and his team hit upon the idea of fixing the genetic lines by using "anther-culture," a technique that involves doubling the genetic complement of pollen grains (the equivalent of sperm) to produce a fixed and highly fertile line, with two identical sets of genetic material. Anther-culture had several advantages over conventional breeding: lines could be fixed in only two years, the process solved some of the fertility problems in the first-generation progeny, and it allowed Jones to recover recombinant lines—those that combined features from both parents—the primary goal of the project.

In 1994 Nerica was unveiled by the WARDA and was made available for community-based field testing in Cote d'Ivoire in 1996. The organization allowed sustenance farmers to choose up to five Nerica varieties, which they would cultivate and then distribute to other community members over the course of the three-year experiment. At the end participants were permitted to buy more Nerica seeds if they chose to—on the grounds that poor farmers would spend their own money only on seeds that had really worked for them. Similar experiments followed in Ghana, Guinea, and Togo; other countries, including Sierra Leone and Nigeria, were soon invited to conduct their own testing. The rice was almost universally hailed by participants for its ability to thrive in difficult conditions, its three-month harvest time (half the time of traditional varieties), its potential to increase crop yields by up to 250 percent, and its high protein content.

In 2002 Jones was named executive secretary of the Ghana-based Forum for Agricultural Research (FARA). In one of his last official acts before leaving the WARDA, Jones helped establish the African Rice Initiative (ARI), which serves as a channel for the distribution of Nerica in several countries across Sub-Saharan Africa. In 2004, which the Food and Agriculture Organization of the United Nations declared the International Year of Rice, Jones was named the corecipient of the World Food Prize. (He shared the honor with a fellow rice scientist from China, Yuan Longping.) He was the first person of African descent to win the award, which is bestowed annually on individuals who have made significant contributions toward improving the quality, quantity, or availability of food in the world.

In 2007, *Time* featured Jones as one of the "100 most influential people in the world." His other accolades include the National Order of Merit of Cote d'Ivoire (2001), the Insignia of the Grand Officer of the Order of the Rokel from the president of Sierra Leone (2004), and the Niigata International Food Award (2010). The WARDA has established an annual lecture series in his honor. Jones, among other roles, is a fellow of the African Academy of Sciences, a board member of the Alliance for a Green Revolution in Africa, and chair of the Global Forum for Agricultural Research. He lives in Accra, Ghana, with his wife and daughter.

There are more than 3,000 varieties of rice within the Nerica species now being used across Africa, forcing global observers to rethink the continent's "will and capacity to solve its [own] problems," as Savitri Mohapatra wrote for the *East Africa Standard* (July 4, 2007). "When Africa breaks free from the grip of poverty and famine—as it now looks poised to do—Jones . . . will have played a pivotal role," Jeffrey Sachs wrote for *Time* (May 14, 2007). "His efforts in creating Nerica are legendary. His work has helped inspire other groups to support an African Green Revolution. The payoff will be a healthy, well-nourished continent on the path of economic growth."

Further Reading

"Eureka for NERICA!" *New Agriculturist*. WREN*media*, July 2004. Web. 13 Aug. 2012.

Sachs, Jeffrey. "The TIME 100: Monty Jones." *TIME Magazine*. Time, 3 May 2007. Web. 13 Aug. 2012.

"Scientist Wins Prize for New African Rice." *Africa Renewal* 18.4 (Jan. 2005): 9. Print.

Kenyon, Cynthia

American biochemist and biotechnology executive

Born: February 18, 1954; Chicago, Illinois

The work of the geneticist Cynthia Kenyon has helped to redefine the way that scientists understand the process of aging. "Once upon a time the study of aging was a scientific career killer right up there with cold fusion," Steven Kotler wrote in *Discover* (November 2004). "Everybody had pretty much agreed that decrepitude was the result of entropy—a seemingly inevitable increase of biological disorder. Then Cynthia Kenyon came along." Kenyon's studies of the microscopic roundworm *Caenorhabditis elegans* (known as *C. elegans*) surprised the scientific community by proving that the aging process is determined by genes, and that it is far more plastic than previously assumed; researchers in her lab have expanded the life spans of the worms they have studied by a factor of six. "People have always thought that, like a car, our body parts eventually wear out," Kenyon told Alex Crevar for *Georgia Magazine* (June 2004), a publication of the University of Georgia. "But we found that over time, when one gene was manipulated, the worm actually remained youthful—in all ways—so that age-related diseases were also postponed." Kenyon's research may prove useful in the treatment of diseases associated with aging in humans and, in 2000, she cofounded Elixir Pharmaceuticals, a company whose aim is to develop drug therapies for such diseases. Kenyon is the Herbert Boyer Distinguished Professor of Biochemistry and Biophysics, and director of the Hillblom Center for the Biology of Aging, at the University of California, San Francisco (UCSF). She has been the principal author of more than seventy publications in such prestigious scientific journals as *Cell*, *Nature*, and *Science*. "Kenyon is part of an elite core of researchers," Andy Evangelista wrote in *UCSF Magazine* (May 2003), "an internationally recognized pioneer who has helped thrust the science of aging to the research forefront."

Early Life and Education

The oldest of three children, Cynthia Kenyon was born on February 18, 1954, in Chicago, Illinois, where her father, James Kenyon, was attending graduate school at the University of Chicago. The family left Chicago after James Kenyon graduated, living briefly in New Jersey before settling in Athens, Georgia. Kenyon's father was appointed to the faculty in the Department of Geography at the University of Georgia; her mother, Jane Kenyon, accepted a position as an administrator in the Department of Physics. Kenyon's parents, as intellectuals and educators, encouraged their daughter's varied interests and natural curiosity. "Cynthia would keep a praying mantis on a leash and feed it from a toothpick dipped in honey," Jane Kenyon told Alex Crevar. "She liked to be outside and would go down to the river and just read all day long. She was incredibly inquisitive, and since she came from a family of teachers we never asked any of our children . . . a question that ended in a yes or no answer."

As a girl, Kenyon took up the French horn, and by high school she had set her heart on a career as an orchestra musician. "I didn't know it at the time, but I was terrible at the French horn. I had almost no talent. I would miss notes and people would get mad at me. It was so unpleasant that I finally quit . . . ," Kenyon told Karen Roy for the *Chicago Tribune* (October 26, 1997). "It taught me that what you should try to find is not only something that you like, but something that you're good at. It's not enough to just work hard, you have to have some talent for it."

When Kenyon entered the University of Georgia as an undergraduate, she struggled to choose a major, vacillating among veterinary medicine, mathematics, philosophy, English, and other subjects. "Many, many different things were very interesting to me," Kenyon told *Current Biography*. "Finally, I just gave up and dropped out because I couldn't decide, and I was running out of time." Instead of returning to school for her junior year, Kenyon worked for a while on a farm, which she did not find satisfying. She discovered her calling when her mother brought home a copy of the book *Molecular Biology of the Gene* by James Watson, one of the scientists who had discovered the double helix structure of DNA. "I looked at it, and I thought:

This is really cool, you know, genes getting switched on and off. And I thought: I'll study that. I loved the idea that biology was logical," Kenyon told David Ewing Duncan for *Discover* (March 2004). "A big tree seemed even more beautiful to me when I imagined thousands of tiny photosynthesis machines inside every leaf." Kenyon returned to school with renewed enthusiasm to pursue degrees in chemistry and biochemistry. "I realized that biology was sensible, that we could understand it the same way we understand electronic circuits," she told Karen Roy. "I was intrigued. This offered a way to learn about life."

After graduating—and becoming class valedictorian—in 1976, Kenyon was awarded a National Science Foundation graduate fellowship, which allowed her to pursue her doctorate in biology at the Massachusetts Institute of Technology (MIT) in Cambridge. There, she worked on a project whose purpose was to examine the response of *E. coli* bacteria to DNA-damaging agents. "That got me interested in development. You took an *E. coli* cell and you treated it with a DNA-damaging agent and it turned into a whole different animal," Kenyon told a reporter for the *New Scientist* (October 18, 2003). Kenyon was first exposed to *C. elegans* in the laboratory next to hers—that of H. Robert Horvitz, who won the 2002 Nobel Prize in physiology and medicine. "In *Caenorhabditis elegans* you have only a thousand cells and people knew where they all came from in the cell lineage, starting from the fertilized egg. It was the most amazing thing I'd seen, almost like the crystal structure of an animal . . . ," Kenyon told the *New Scientist* reporter. "And I thought, wow, that's what I want to do."

C. elegans, also known as nematodes or roundworms, are ideal for studying genes because of their short life spans; they live an average of two to three weeks, reaching maturity in just three days. The nematode's transparent body also makes it easy to study its cellular structure under a microscope. Additionally, many processes seen in nematodes have been duplicated in humans. Humans share at least 50 percent of the genes found in nematodes, including the genes that govern cell division, cell migration, cell differentiation, and tissue-pattern formation. "Everything else we know about the worm, we find happens in a similar way in humans," Kenyon told Karen Roy. "Muscles are made the same way, cells divide the same way, the body plan is set up with the same mechanism. Everything else that we thought would be different is the same."

Life's Work

Kenyon earned her PhD degree in 1981. The following year she began working with *C. elegans*, in the course of her postdoctoral work at the Medical Research Council Laboratory of Molecular Biology in Cambridge, England. She trained there as a developmental biologist under the Nobel laureate Sydney Bremmer, who was studying the roles of genes in the development of nematodes. It was there that Kenyon developed an interest in the subject of aging. One day she noticed an old lab dish containing sterile worms that she had forgotten to throw out. "There were all these old worms on the plate. I had never seen an old worm. I had never even thought about an old worm," Kenyon told Nell Boyce for *U.S. News & World Report* (December 29, 2003). "I felt sorry for them. And I felt sorry for myself, 'Oh, I'm getting old, too.' And right on the heels of that, I thought, 'Oh, my gosh, you could study this.'"

In 1986 Kenyon accepted a position as an assistant professor at UCSF. Soon afterward she began to study aging in *C. elegans*. At that time the scientific community viewed the field of aging with disdain, and for that reason she forbade her colleagues from using the word "aging" on grant applications. "There was a lot of science done in that area that was not of a very high quality," Kenyon told *Current Biography*. "Plus there was a general feeling among molecular biologists that there was nothing to study. For example, one person who I worked with said, 'Be careful, Cynthia, you might fall off the end of the earth.'" Still, Kenyon believed that there was more to aging than an inevitable descent into entropy. "When I looked at it, I could see that different animals have different life spans," Kenyon told Stephen S. Hall for *Smithsonian* (March 2004). "The mouse lives two years and the bat can live fifty years and the canary lives about fifteen or so years. They're all small animals. They're warmblooded. And they're not that different, really, in such a fundamental way from each other."

Convinced that genes controlled the rates and effects of aging, Kenyon searched for proof of her hypotheses in *C. elegans*. "There was already a mutant worm that was reported to live 50 percent longer. It had been isolated by Michael [Klass] ten years earlier and been studied by Tom Johnson's lab [at the University of Colorado, Boulder]," Kenyon explained to Steven Kotler. "They thought maybe the mutant lived long because it didn't eat well or didn't reproduce well, but I thought there's a real set of dedicated genes for aging and that was the reason why. So we looked for these mutants."

Kenyon and her team shook up the scientific community when they announced in the December 1993 issue of *Nature* that they had been able—through the manipulation of a single gene, daf-2—to double the nematode's life span. The daf-2 gene works as a sort of brake on the function of another gene, daf-16. When the nematode is faced with overcrowding or a food shortage, the daf-16 gene causes nematode larvae to enter what is known as the dauer state—in which the larva's development is halted and it enters a period of stasis, allowing it to live for extended periods of time without any food or water. Scientists were already aware that daf-2 and daf-16 affected the dauer state; Kenyon and her team discovered that these genes also influenced aging, and that weakening daf-2 in an adult worm would double its life span. Not only did the daf-2 mutant worms live longer, but their aging processes slowed; the worms remained healthy until a few days before their deaths. A comparable development in humans would be a person's living to be 150 and having the body of a forty-five-year-old at the age of ninety. "After two weeks, the normal worm is basically in a nursing home, or dead already. It's lying quietly, it looks old," Kenyon told Karen Roy. "In contrast, the altered worm is still active. It isn't just hanging onto life, it's out on the tennis court."

Though the results had yet to be replicated in mammals, Kenyon's discovery suggested that it might be possible to retard aging in humans as well. The daf-2 gene encodes a hormone receptor, or a protein that allows tissues to respond to hormones. Kenyon's lab found that mutations that lower the activity of the receptor, making the tissues less responsive to the hormone, also increase life spans. "Our result told us that the normal function of these hormones was to speed up aging . . . ," Kenyon told Steven Kotler. "The same kinds of hormones are found in all animals. In people, it's insulin, which is used in food utilization, and a hormone called IGF-1, insulin-like growth factor." As a direct result of Kenyon's research, other scientists later determined that the same genes also regulate the aging process in flies and mice—suggesting that the mechanisms that govern aging may be universal in all species.

Philip Anderson, a professor of genetics at the University of Wisconsin–Madison, urged caution before applying those results to humans, but he also told Malcolm Ritter of the Associated Press (December 1, 1993), "When we understand how genes like daf-2 and daf-16 function, we will I think have insights into how aging works in humans, and perhaps insights that are directly applicable."

Understanding the Mechanism of Aging

In October 1997, Kenyon was appointed the Herbert Boyer Distinguished Professor of Biochemistry and Biophysics at UCSF. The following year, she made another important breakthrough in understanding the mechanism of aging. In an article published in *Cell*, Kenyon and her associates announced their finding that the aging process does not occur in individual cells but is coordinated by signals exchanged by the different tissues. "Our study indicates there is a mechanism that causes all of the cells in the animal to reach a consensus," Kenyon told a reporter for *Gene Therapy Weekly* (November 2, 1998). "And that mechanism appears to be sparked into action by particular genes acting within certain types of cells." Using what is known as "mosaic" analysis, Kenyon and other researchers in her lab diminished or eliminated the daf-2 gene—and later, the daf-16 gene—in a variety of cells to see which affected the organism as a whole. They determined that a change in certain groups of cells, such as nerve or (as a later study indicated) intestinal cells, can affect the aging process of all cells in an organism, even those that contain the normal daf-2 or daf-16 gene. "In adult animals, by acting in signaling cells to control the production of a second signal, the daf-2 gene may be able to coordinate the aging process of all the cells in the animal, so that they all age together at the same rate," Kenyon told the reporter for *Gene Therapy Weekly*.

In 1999 Kenyon and her colleague Honor Hsin discovered a second hormonal pathway—in addition to the insulin receptor encoded by daf-2—that affects the aging process. By using a laser microbeam to remove the reproductive cells, also known as germ cells, Kenyon and Hsin extended the *C. elegans*'s life span by 60 percent. "It turns out that the germ cells do two things: they produce the next generation, the sperm and the eggs—they are the next generation in a sense—and in

> **"A big tree seemed even more beautiful to me when I imagined thousands of tiny photosynthesis machines inside every leaf."**

addition they speed up the aging of the body," Kenyon told *Current Biography*. "Let's suppose a mutation took place, or something happened to slow down the development of the germ line. What would happen? The animal would reproduce later. It would take it longer to be able to reproduce, but if you did this, the germ cells would no longer be able to speed up aging as well, and as a consequence the animal would age more slowly. . . . So it's a way of bringing the timing of reproduction into register with the aging of the body, allowing animals to reproduce in their prime, which is what you want. So my prediction is that this will turn out to be important in evolution, in the evolution of life strategies for different organisms—how fast it develops, when to reproduce."

Kenyon believes that it may be possible, once the mechanism in germ cells that controls aging has been identified, to manipulate that mechanism so as to extend the life spans of nematodes without also rendering them infertile. Nematodes whose life spans have been extended only through the manipulation of daf-2 remain fertile. Kenyon disputes the prevailing notion that life spans can be extended only at the cost of reduced fertility or a lower metabolism. "People think that if you extend life span there has to be a trade-off, but how could that be true?" she said to Alex Crevar. "Think about it, evolution started with this little primitive animal that is one of your ancestors and that gave rise to *C. elegans* and then to some animal that gave rise to you. If every time you changed genes there was a trade-off, we would never be so much superior. We are amazing. If every time there was some downside, we would not live 1,000 times longer than the nematode worm . . . and we are not 1,000 times worse off."

In another research project, conducted with Javier Apfeld, Kenyon discovered that sensory perception also influences the aging process. Sensory neurons in the nematodes' heads and tails gather information about their environment, such as temperature, movement, and the presence of food. Kenyon and Apfeld found that genetic mutations that cause defects in the neurons, inhibiting sensory signals, also extend life span. "The signal affecting lifespan may be a substance that the animal can smell or taste," Apfeld and Kenyon wrote in a paper published in *Nature* (December 16, 1999). "For example, it may be a pheromone such as dauer pheromone, which reflects population density, or a compound that originates from organic material and reflects food availability." In a later study, however, Kenyon determined that the signal affecting life span was not the dauer pheromone. The results of that study, published in *Neuron* in 2004, also revealed that taking away either of two chemosensory receptors—that is, proteins that stick through the tips of the neurons and bind to odorants or substances that the worms taste—extends life span as well.

In 2003 Kenyon published a study in *Science* (May 16, 2003), addressing the question of why the elderly become susceptible to age-related diseases. For example, when the gene that causes Huntington's disease is introduced into a worm, the animal contracts Huntington's disease as it ages. By contrast, the worms with manipulated daf-2 genes also get Huntington's disease, but not until they are older. Kenyon's lab found that those worms produced a greater number of "small heat shock proteins," also known as chaperones, which attach themselves to unfolded cellular proteins and prevent them from forming harmful aggregations that can lead to many age-related diseases, such as Alzheimer's and Parkinson's. "The chaperones also see to it that a damaged protein is either fixed or it's put into the wastebasket and degraded, making room for another protein," Kenyon told *Current Biography*. "We found that these chaperones were able to do two things: to increase life span and to delay the time of onset of Huntington's disease. So by keeping the cell's normal proteins, and also the Huntington protein, from misfolding, these chaperones may couple the normal aging process to the time of onset of this disease."

Later that year Kenyon published another study in *Nature* (June 29, 2003). Working with both normal and daf-2 mutant nematodes, researchers in Kenyon's lab identified a large number of genes, including antioxidant, chaperone, antimicrobial, and metabolic genes, that were either more or less active in the long-lived mutants. They then used a technique known as RNA interference to disable these genes one at a time. While manipulating some genes increased life spans by only 10 to 30 percent, manipulating the daf-2 gene can mobilize them all at once, producing a huge increase in life span. "The daf-2 gene acts as an orchestra conductor," Kenyon told Alex Crevar. "It's a very powerful regulator that brings together all the instruments in the orchestra. The cellos, for instance, are the antioxidant genes, the piccolos prevent infection, and the French horns are involved in things like fat transport. When we change the way one responds to a hormone, we know that about 100 genes turn up or down—or on or off—for a large cumulative effect."

In October 2004, the journal *Science* reported that by combining techniques—weakening the daf-2 gene and destroying cells in the reproductive system—Kenyon, along with her colleagues Nuno Arantes-Oliveria and Jennifer Berman, had extended the lifetime of the nematode sixfold. "The thing that was so remarkable about those worms was how healthy they were. You might think they would just lie there or something—but they don't. They move around," Kenyon told *Current Biography*. "They're very healthy and very active, even when they're quite old. . . . And they look good. They're not skinny or anything. They're nice-looking worms." The equivalent in humans would be living healthy, active lives for 500 years.

Her findings notwithstanding, Kenyon does not predict that human life spans will increase dramatically in the near future. She believes that the benefits of this research will be seen, instead, in treatments for age-related diseases, such as type II diabetes, obesity, cancer, and heart disease, and perhaps in protein-aggregation disorders such as

Huntington's disease. "There are lots of different age-related diseases that seem to be postponed in these long-lived animals, whether it's worms or flies or mice. . . . There's something about being biologically or physiologically older that makes you more susceptible," Kenyon told *Current Biography*. "So people traditionally have been going after this disease or that disease—cancer, osteoporosis—separately, but if you could go after aging, which is the biggest risk factor for the vast majority of all these diseases, then you might be able to delay the time of onset for many all at once."

Around 1996 Kenyon began courting venture capitalists, showing them a five-minute film about the nematodes she had studied. "I show them the movie, of worms that are normal that are about to die after two weeks, and the altered worms that are still moving," she told Stephen S. Hall. "They see it with their eyes, you know? . . . And that's really all you have to do. It speaks for itself." In 2000 Kenyon and Leonard Guarente, the Novartis Professor of Biology at MIT, founded Elixir Pharmaceuticals, a Boston-based company dedicated to developing ways of extending youthfulness and improving the quality of old age. Elixir finalized an agreement with UCSF in 2003 to access discoveries made by Kenyon and her laboratory for potential drug development. "We may not have a perfect drug in ten years, but we'll have something to build on. Potentially, we could have it sooner," Kenyon told Stephen S. Hall.

Kenyon's ability to explain science in ways that are accessible to most laypeople has made her a popular spokesperson in the media. She has appeared on the Discovery Channel and ABC News, was featured on a Scientific American Frontiers TV special, and has also been heard on radio specials for National Public Radio and the BBC. She is currently working on a book about aging for a general readership.

In April 2003 Kenyon was one of six speakers at the fiftieth-anniversary celebration of the discovery of DNA, held in Cambridge, England, at which James Watson was also present. Kenyon is a member of the American Academy of Arts and Sciences, a member of the National Academy of the Sciences, and an Ellison Medical Foundation Senior Scholar. She has been awarded the Wiersma Visiting Professorship at the California Institute of Technology, a Searle Scholarship, a

Packard Fellowship, the Discover Award for Innovation in Basic Research, the Award for Distinguished Research in the Biomedical Sciences from the Association of American Medical Colleges, and the King Faisal International Prize for Medicine. She served as the president of the Genetics Society of America in 2004.

Personal Life

In her leisure time, Kenyon still plays the French horn—as well as the guitar and Alpine horn. Taking a cue from her own research, she maintains a diet that keeps her insulin levels low. "I don't want to get old. And I don't think I'm the only person that feels that way," she told Karen Lurie for *ScienCentral News* (March 26, 2004). "In fact, if you read Shakespeare's sonnets, so many of them are about the anguish of aging. People don't like to get old, they don't like to lose their abilities, their capacities. So for people who love life, like I do, what could be better?"

Further Reading

Block, Melissa. "Discovering the Genetic Controls That Dictate Life Span" *Life Extension Magazine* June 2002: 51. Print.

Evangelista, Andy. "Cynthia Kenyon: Probing the Prospects of Perpetual Youth." *UCSF Magazine*. The Regents of the University of California. May 2003, Web. 9 Aug. 2012.

Kotler, Steven. "Basic Research: Cynthia Kenyon." *Discover Magazine* Nov. 2004. Print.

Roy, Karen. "A Different Tune." *Chicago Tribune* 26 Oct. 1997: WomanNews 3. Print.

Kiessling, Laura

American biochemist

Born: September 21, 1960; Milwaukee, Wisconsin

Biochemist Laura L. Kiessling has conducted groundbreaking research into the causes and treatment of inflammation of body tissue. She has also pioneered the study of proteins that may cause the brain lesions associated with Alzheimer's disease. Kiessling's work has garnered her several prestigious awards, including a Beckman Young Investigator award and a MacArthur fellowship. She is a professor of chemistry and biochemistry at the University of Wisconsin–Madison.

Early Life and Education

Born on September 21, 1960, in Milwaukee, Wisconsin, to William E. and LaVonne (Korth) Kiessling, Laura Lee Kiessling began performing scientific experiments at an early age. "No one in my family is a scientist," she told Alisa Zapp Machalek in an interview for *Findings* (February 2001). "But my brother Bill and I used to like to do experiments when we were little." The two started out with a basic experiment kit, which came with instructions for making a lie detector and other items. Soon they began using the kit to carry out experiments of their own devising: "We rigged it up to give a small shock to the doorknob and then tricked our younger brother Mark into opening the door." Not all of her experiments succeeded, as she revealed to Machalek. "One time I tried to put little motors on the end of a Frisbee that would fire in opposite directions to make it spin really fast," she said. Recalling that the Frisbee would not fly, she added, "I forgot about the idea of buoyancy. Let's just say it wasn't very aerodynamic."

Kiessling received a bachelor's degree in chemistry from the Massachusetts Institute of Technology, in Cambridge, Massachusetts, in 1983. Six years later she completed a PhD degree in chemistry at Yale University in New Haven, Connecticut. While there, she served as a teaching assistant and later as a research assistant. After graduating from Yale, Kiessling spent two years as a research fellow at the California Institute of Technology in Pasadena. In 1991 she accepted a position as assistant professor of chemistry at the University of Wisconsin. Six years later she was promoted to associate professor, and in 1999 she became a full professor of chemistry and biochemistry.

Life's Work

During her early years at the University of Wisconsin, Kiessling began to work on a way to fight inflammation, which is one of the body's reactions to injury or disease. To prevent the spread of the condition and allow the injured area to heal, the body creates redness, swelling,

> "The collaboration is very tight. On a daily basis, my students aren't only talking to me, but the chemists and the biochemists are talking to each other."

heat, and pain in the affected area. This reaction occurs when white blood cells leave the bloodstream and enter the tissues surrounding the injury. When they receive the signal to exit the bloodstream, the cells, which are coated in carbohydrates, move toward the walls of the blood vessels within which they are traveling. Because these walls are covered with sticky proteins, the carbohydrate-coated white blood cells adhere to the walls, a process that causes the blood vessels to widen so that the cells can travel to the injured tissues. Although inflammation is typically associated with health and healing, it can also be dangerous, particularly when infected white blood cells spread disease. Kiessling sought to combat this danger by studying white blood cells at the molecular rather than the cellular level. She hoped to create a synthetic carbohydrate that would stick to the proteins covering the blood-vessel walls in such a way that would block the white blood cells from leaving the vessels and stop the process that would lead to inflammation. Kiessling continued her research throughout the 1990s. In a 1998 article published in the journal *Nature*, she announced that she had created just such a substance. In addition to blocking white blood cells from leaving the bloodstream by sticking to the walls, the synthetic carbohydrates snipped those proteins from the walls, thereby preventing any further spread of diseased blood cells. A year after her announcement, Kiessling received a $285,000 "genius" grant from the John D. and Catherine T. MacArthur Foundation.

Kiessling has also conducted innovative research on Alzheimer's disease. In 1997 she and Regina Murphy, a chemical engineer and fellow professor at the University of Wisconsin, announced that they had discovered a way to prevent the formation of the brain lesions associated with Alzheimer's disease. The brains of Alzheimer's patients contain plaque deposits, which are made up of sticky protein molecules. These proteins adhere to each other and, together with dead and dying cells, form clumps between brain nerve cells. Kiessling and Murphy wanted to find a way to disrupt the aggregations of these sticky proteins, thus preventing the creation of the damaging plaque. To that end, they created synthetic "inhibitor molecules," which bind with the proteins and stop them from hardening into plaque. They hope eventually to develop the synthetic molecules into a substance that can be used to treat Alzheimer's.

In 2002, Kiessling and a graduate student, Jason Gestwicki, published an article in *Nature* detailing their discovery that clusters of receptors—which receive signals from the environment—on the surface membranes of disease-causing *E. coli* bacteria communicate with each other about how and where to move. This finding could help researchers design compounds that will interfere with the signaling process, which could help protect people from disease.

Kiessling is known as an excellent teacher as well as a first-rate scientist. She has brought together her chemistry and biochemistry students by having the former fabricate molecules and having the latter study how those molecules behave in living cells. "What I like is that the chemistry students really get a good understanding of the biochemistry experiments and the biochemistry students come away with a much better understanding of what compounds are possible to make," she told Machalek. "The collaboration is very tight. On a daily basis, my students aren't only talking to me, but the chemists and the biochemists are talking to each other."

In addition to the MacArthur fellowship and Beckman award, Kiessling has received the Arthur C. Cope Scholar Award, the Alfred P. Sloan Foundation Fellowship, the National Science Foundation's National Young Investigator Award, and the Dreyfus Teacher-Scholar Award. In 2008, Kiessling was named a Guggenheim Fellow. She has served on the editorial boards of *Chemistry and Biology* and *Organic Reactions*, and is a member of the American Chemical Society, the American Association for the Advancement of Science, the Society for Glycobiology, and the American Society for Biochemistry and Molecular Biology.

Kiessling is married to Ron Raines, a biochemist at the University of Wisconsin. They have a daughter, Kyra, whose name was derived by combining the first two letters of each of her parents' last names. In her spare time, Kiessling enjoys canoeing, rowing, and running.

Further Reading

Johnson, Mark. "Persistence, Genius Mix for Chemist." *JSOnline*. Milwaukee Journal Sentinel, 30 Dec. 2007. Web. 9 Aug. 2012.

Mader, Becca. "Shaw Scientist Grants Provide Support To Promising Researchers." *The Business Journal* 26 Apr. 2002: 21. Print.

Sakai, Jill. "Two Faculty Elected to National Academy of Sciences." *University of Wisconsin-Madison News.* Board of Regents of the University of Wisconsin System, 1 May 2007. Web. 9 Aug. 2012.

Moncada, Salvador

Honduran pharmacologist

Born: December 3, 1944; Tegucigalpa, Honduras

To the general public, the most visible by-product of the research of the Honduran-born pharmacologist Salvador Moncada is Viagra, the drug marketed in 1998 by the company Pfizer to treat "erectile dysfunction." The development of Viagra is directly related to Moncada's groundbreaking cardiovascular research into nitric oxide (NO), a molecule—previously perceived as a mere pollutant—that he proved was responsible for a bevy of functions inside the human body. The discovery of the significance of NO was deemed important enough that the 1998 Nobel Prize in medicine was awarded to three scientists who had conducted research on it, but Moncada was not among them. In the view of many scientists, Moncada's contributions to the NO-related discoveries rivaled those of the prizewinners; his omission consequently became a major controversy that damaged the credibility of the Nobel Prize.

Early Life and Education

Salvador Moncada was born in Tegucigalpa, Honduras, on December 3, 1944. His mother's side of the family, of Jewish descent, had fled from the German annexation of Austria in 1938. Moncada's father, a native Honduran, fled the country and its dictatorship in 1948. Moncada enrolled at the University of El Salvador in 1962, graduating in 1970 with a master's degree in medicine and surgery. He found a teaching position at his alma mater, but it proved to be short-lived; because Moncada had participated in demonstrations against the right-wing dictatorship governing the country, agents of the government tortured him and deported him back to Honduras, forbidding his wife and child to leave with him. Moncada attempted to find work in medical research in the United States but was denied a visa, as his political convictions had branded him a radical. Instead, he arrived in

London in 1971 to work at the Royal College of Surgeons under Dr. John Vane. During Moncada's first several years at the Royal College, he contributed to research on the mechanism that allows aspirin to act as a painkiller and anti-inflammatory agent. He published several significant papers on the subject and earned his PhD in 1973.

Life's Work

The following year he traveled to Honduras to look for a medical research position but found nothing. He returned to England in 1975 and rejoined Vane, who had since left the Royal College for Wellcome Research Laboratories, located in Beckenham, Kent. In his first year there, Moncada conducted research that contributed to the discovery of thromboxane synthase (a compound linked to a variety of conditions including heart failure) and ways to inhibit its production. Vane, Moncada, and their team made another major breakthrough in 1976, when they discovered prostacyclin, a hormone created in the inner lining of blood vessels that prevents platelets (cells that cause blood clotting) from sticking to vessel walls. Prostacyclin was later employed in the most widely-used treatment for a condition called primary pulmonary hypertension, a rare lung disorder. (The discovery of prostacyclin was one of the crowning achievements of the career of Vane, who won a Nobel Prize in 1982.) Moncada was promoted in 1984, becoming the director of the Therapeutic Research division at Wellcome. Two years later he became the director of all UK-based research for the company. Under Moncada's direction Wellcome, scientists developed treatments for several medical conditions, including epilepsy, migraine headaches, and malaria.

Moncada's investigations into the properties of blood vessel lining continued as he researched what was then known as the "endothelial-derived relaxing factor," or EDRF. In 1980 Dr. Robert Furchgott of the Downstate Medical Center in Brooklyn, New York (affiliated with the State University of New York), found that unknown agents made by endothelial cells (a type of cell found throughout the human body) line the blood vessels and cause muscles to relax, thus lowering blood pressure. However, the fleeting nature of the compound—EDRF has a life span of only five seconds—made it difficult to identify. Moncada and his team joined a number of scientists attempting to identify EDRF. They discovered that it bore a strong resemblance to "free radicals," or relatively unstable molecules that "steal" electrons from nearby stable ones.

Around the same time, other scientists were studying the effects of nitric oxide (NO) on the human body. NO, released in automobile exhaust, is a primary ingredient in smog. However, studies had found that the incidence of nitric oxide in the human body was higher than could be accounted for by pollution alone. In fact, one investigation, conducted by John B. Hibbs Jr. of the Utah School of Medicine, found that NO was present in macrophages, a type of immune-system cell. Hibbs speculated that the molecule might prove useful in fighting infections or tumors. In 1986 Robert Furchgott and another pharmacologist, Louis Ignarro of the University of California at Los Angeles, theorized that, because NO is a free radical, and because Moncada had demonstrated that EDRF behaved like a free radical, the two compounds might be the same. While the theory was shunned by most biochemists, Moncada jumped on the idea. At Wellcome, Moncada adapted a machine used to measure nitric oxide in car exhaust to detect nitric oxide emissions from human cells. He succeeded in finding the molecule. Next, he exposed muscle cells to pure nitric oxide, and they relaxed, thus proving that nitric oxide was indeed EDRF. Moncada's team also found that NO is produced by an enzyme called nitric oxide synthase (NOS), which is activated by a rise in the calcium levels in endothelial cells. NOS converts a common amino acid, L-arginine, into NO.

In 1989 Moncada invited the international scientific community to the Royal Society (of which he had become a member the previous year) to discuss his discovery. "I had to beg people to come," he told Randi Epstein for the *London Guardian* (November 17, 1992). "People wrote to me saying, 'You must be mad. Nitric oxide is a pollutant.'" Over the next few years, though, the scientific community recognized the importance of Moncada's discovery, which was also touted by major newspapers as scientists began to understand the wide-ranging properties of NO: Scientists found, for example, that NO blocks cholesterol plaque, a substance which can clog coronary arteries. That discovery in turn solved another scientific mystery: why nitroglycerine was effective in easing angina pectoris, a coronary condition. (Moncada's discoveries demonstrated that nitroglycerin was converted into nitric oxide in blood tissues.) Moncada's team also found that the brain produces NO in the same way as the blood vessels, and that NO is used as a neurotransmitter—a message carrier in the brain. Other scientists found that the overproduction of NO can lead to strokes or Huntington's disease. In 1991, a team headed by K. E. Andersson of Lund University Hospital in Sweden found that NO plays a crucial role in the biochemical process leading to an erection: As the team explained, sexual arousal in the brain prompts the release of NO in the nerves of the corpus cavernosum, a muscle in the penis. The NO, in turn, causes the muscle to relax, allowing blood to enter the tissues in the area, thus leading to an erection. (Viagra, the popular drug introduced by the drug company Pfizer in 1998, blocks an enzyme that prevents the spread of NO in the penis.) Moncada's team even found that nitric oxide was released in starfish stomachs. "We have known for years that we excrete nitrates in our urine, but no one has ever really thought about that before.... We were always thinking in terms of complex mediators," Moncada explained to Gina Kolata for the *New York Times* (July 2, 1991). "But this is very simple. We are beginning to realize that it is so simple it must be universal. And it is so beautiful that it must be in many different biological systems." As he further explained to Nina Hall for the *Observer* (April 11, 1994), "we were not prepared for the idea that a gas which is released like a little puff of perfume from an atomizer could mediate a response. It has made us question some of the basic tenets of biology." The numerous discoveries regarding NO led the magazine *Science* to name it "Molecule of the Year" for 1992, and, as the excitement surrounding the discoveries grew, speculation abounded that Moncada was due for a Nobel Prize in medicine.

The Wolfson Institute for Biomedical Research

Moncada's discoveries had made him an extremely valuable asset to his employer, Wellcome Research Laboratories. But his professional relationship became strained in 1995, when the British pharmaceutical giant Glaxo staged a hostile takeover of Wellcome. (The merger made the new company, Glaxo Wellcome, the largest pharmaceutical company in the world.) Moncada joined a chorus of other Wellcome employees in expressing concern over likely budget cuts and staff reductions. In August 1995, after the takeover, Moncada announced his departure from Wellcome. Several weeks later, he announced that he was going to spearhead the development of a new academic research institute, the Wolfson Institute for Biomedical Research, based at the University College London. It was to be funded by money from the government, private foundations, and Glaxo. "Our mission is to combine cutting edge research with a focus on delivering tangible outcomes in the quest for new drugs," he told a reporter for *Pharmacy Today* (November 1995). "It represents an obvious opportunity for investment by the pharmaceutical industry, which is increasingly establishing external collaborations to aid drug discovery research." In return for their financial backing of the institute, Glaxo retained some commercial rights on the research at Wolfson. Moncada, as the institute's scientific director, recruited several Wellcome colleagues as well as research teams from St. George's Hospital and King's College medical schools, and declared the institute would concentrate on cardiovascular, neurological, and cancer research. He gathered experts specializing in a range of different fields in order to create an environment that would encourage intellectual cross-fertilization and unorthodox thinking. "Very often great ideas are written on the back of an envelope in a bar," Moncada told Michael Hanson for the *London Express* (April 11, 2000). "To make discoveries in science you have to let the mind roam freely." In February 1996 the Wolfson Institute opened temporary quarters at the Rayne Institute in London. (It relocated to its permanent address, the Cruciform building in the University College London campus, in April 2000.)

Several prominent members of the scientific community expressed surprise when, in October 1996, the annual Albert Lasker Basic Medical Research Award, regarded by many as a precursor to the Nobel Prize, was awarded to Robert Furchgott and Ferid Murad, but not to Moncada. (Furchgott and Murad were pioneers in nitric oxide research.) John Vane led a group of dissenters who maintained that, because Moncada was the scientist who proved that EDRF was NO, he was entitled to any awards given in relation to the discovery. Vane and his colleagues' complaints were published in an article in the journal *Science* on October 11, 1996. Later that month, however, both Furchgott and Moncada were the recipients of the prestigious Gregory Pincus Medal, awarded by the Worcester Foundation for Biomedical Research. At the awards symposium Moncada told the press, as quoted by Elaine Thompson for the *Massachusetts Telegram & Gazette*

> **"Very often great ideas are written on the back of an envelope in a bar."**

(October 24, 1996): "I don't think scientists work for awards. If they come, one is greatly happy. If they don't, the satisfaction of doing the work is enough."

On October 12, 1998, Furchgott, Murad, and Louis Ignarro were announced as winners of the Nobel Prize for Medicine; Moncada had once again been left out. While there was no official limit on the number of corecipients for the Lasker award—the award has been given to as many as five people at the same time—the limit on corecipients for the Nobel Prize is three. Again, a number of biomedical scientists protested the omission. Publicly, Moncada congratulated the winners, but newspaper reporters quoted accounts and rumors that he was disappointed and bitter. According to Nell Boyce and Andy Coghlan for *New Scientist* (October 17, 1998), Moncada stated, "I'm not angry, just surprised." Furchgott, the only one of the prize recipients whose contributions were universally regarded as instrumental to the nitric oxide discoveries, told Nigel Williams for *Science* (October 23, 1998), "I think very highly of the work of the other winners, but I'm unhappy Salvador is not included." Several academics even declared that Moncada's omission had damaged the credibility of the Nobel. By December 1998 Moncada had abandoned his conciliatory attitude, declaring that the decision to award of the prize only to United States-based researchers displayed the overwhelming lobbying power of the United States: "It's incomprehensible," a reporter for the Agence France Presse (December 5, 1998) quoted him as telling the Swedish daily newspaper *Svenska Dagbladet*. "In science it is the proof that counts not the hypothesis." Years later, many scientists still comment on Moncada's omission as a mistake, and it has been frequently cited by those who argue that the Nobel selection rules should be amended.

Although Moncada was passed over by the Nobel committee, he later received recognition of another sort. In the mid-1990s an information-gathering group called the Institute for Scientific Information, based in Philadelphia, reported that Moncada was one of the world's most frequently cited scientists in academic research papers. Moncada was the sixth most-cited scientist in papers published from 1990 until 1996, and the most-cited scientist in the field of biomedical research in papers published between January 1990 and June 1997, with 30,081 citations. (That figure represented almost twice the number of citations of the runner-up.) Moncada welcomed the findings, telling reporters they meant more to him than a Nobel. "I prefer not to be part of the prize and to have the quotations," he told Sarah Eldridge for the *London Guardian* (November 20, 2001). "It means my work has been recognized by my peers. The quotes are the cleanest recognition because there's no politics involved. Quotation is the only real honest measure of impact, especially when there's so much politics around prize-giving." In 2003 Thomson ISI, the former Institute for Scientific Information, found that Moncada was the second-most-cited scientist in academic papers over the previous twenty years.

Moncada has been showered with a host of other awards and honors for his work, including the Spanish Prince of Asturias Prize for technical and scientific research (1990); the Amsterdam Prize for Medicine (1992); the Heineken Prize for Medicine (1992); France's Roussel Uclaf prize, which he shared with Furchgott and Ignarro (1994); England's Royal Medal (1994); and the Society of Apothecaries of London Galen Medal (1999). He became a fellow of the Royal College of Physicians in London in 1994, and in the same year became a Foreign Associate of the US National Academy of Sciences. He received honorary doctorates from Edinburgh University in 2000 and the University of Dundee in 2001. He received a knighthood from Queen Elizabeth in the 2010 New Year's Honours for his services to science.

Moncada lives in London. He married Princess Maria-Esmeralda of Belgium, the half-sister of the Belgian King Albert II (her professional name is Esmeralda de Rethy) on April 5, 1998. They have a daughter, Alexandra Leopoldine, and a son, Leopoldo Daniel. Moncada told Sylvain Comeau for the *McGill Reporter* (April 19, 2001) that it is the thrill of discovery that keeps him interested in science: "Sometimes you don't know whether what you are seeking is reality, or whether it is something you are imagining. It makes you very insecure, but it is the same feeling as the explorer of a new area of the world, or of the moon; you are going where nobody has been before."

Further Reading

"All That Pollutes Isn't Bad." *McGill Reporter* 33.15. McGill University, 19 Apr. 2001. Web. 9 Aug. 2012.

Sample, Ian. "The Giants of Science." *The Guardian* 25 Sept. 2003: 4. Print.

Williams, Nigel. "NO New Is Good News-But Only for Three Americans." *Science* 23 Oct. 1998: 610–611. Print.

Olopade, Olufunmilayo

Nigerian-American oncologist

Born: April 29, 1957; Nigeria

Also known as: Funmi Olopade

Assessing the groundbreaking research of the Nigerian-born geneticist and oncologist Olufunmilayo Olopade, James Madara, dean of the Pritzker School of Medicine at the University of Chicago, told a writer for the Healthy Life for All Foundation website, "Big thinking is what translates into big discovery. Dr. Olopade's thinking has been precisely on this scale: a world- and genome-wide view of connections between African descent and one of the most common cancers in women." Over the last two decades Olopade's ability to think in big and small ways as a researcher and physician—bridging intellectual zeal with patient advocacy, scientific inquiry with cultural empathy, medical ethics with social justice—has established her as a leading voice in the international medical community. In the 1990s Olopade launched a series of landmark, cross-cultural studies to identify the nature of gene mutations in breast cancer, particularly in the cases of women of African descent. With support from the National Institutes of Health and the Breast Cancer Research Foundation, Olopade compared the genetic profiles of several hundred women across Africa, North America, and Europe; her research revealed that women of African descent expressed different gene mutations than their Caucasian counterparts and stood at higher risk for more aggressive forms of breast cancer at younger ages. She also found that the gene mutations expressed in African women resisted the conventional therapies prescribed to patients of European descent. Motivated by the cultural and environmental facets of her research, in 1992 Olopade established the Cancer Risk Clinic at the University of Chicago Medical Center to translate her findings into effective clinical practice, addressing the cases of African women whose genetics made them particularly susceptible to disease. Currently, Olopade heads the Center for Clinical Cancer Genetics in the Department of Medicine at the University of Chicago.

In 2005 Olopade received a $500,000 John D. and Catherine T. MacArthur Foundation fellowship—often called the "genius grant"—for her pioneering research into the molecular genetics of breast cancer. The MacArthur award underscored the importance of Olopade's cross-continental, interdisciplinary approach to the prevention, detection, and treatment of breast cancer. It also served as an acknowledgement of Olopade's efforts to apply scientific inquiry to social activism. After receiving the award, Olopade told Charles Storch for the *Chicago Tribune* (September 20, 2005), "People in Africa have been understudied, and even African Americans in this country have been understudied." She will, she added, "continue . . . advocating for the vulnerable."

Early Life and Education

The fifth of six children, Olopade was born Olufunmilayo Falusi on April 29, 1957, in Nigeria, in West Africa. (She is known to friends and colleagues as Funmi.) Details surrounding Olopade's formative years in Nigeria remain scant; it is known, however, that her early interest in medicine was sparked by her parents' concern over the dearth of trained physicians in their village. When she reached her late teens, Olopade enrolled in medical school at the University of Ibadan's College of Medicine, Nigeria's oldest and most prestigious institution of higher learning. As a student she showed exceptional promise, garnering departmental honors in pediatrics, surgery, and clinical sciences. In 1980 Olopade received her MD and was awarded, among other distinctions, the University of Ibadan College of Medicine's Faculty Prize and the Nigerian Medical Association Award for excellence in medicine. After graduation Olopade participated in the National Youth Service Corps—a Nigerian government-sponsored mandatory program for university graduates—serving as a medical officer at the Nigerian Navy Hospital in Lagos. After working for a year in that program, she relocated to the United States to serve an internship and residency in internal medicine at Cook County Hospital in Chicago, Illinois. A year after her arrival, a military coup took place in Nigeria, leading to Olopade's decision to settle in the United States. In 1986 she was promoted to chief resident at the hospital.

Life's Work

In 1987 Olopade accepted a postdoctoral fellowship in hematology and oncology at the University of Chicago. Under the advisement of Janet Rowley, Blum-Riese Distinguished Professor of Medicine, Olopade researched the molecular genetics of cancer, examining the specific link between genetic mutation and susceptibility to the disease. She completed her fellowship in 1991 and accepted a faculty appointment as assistant professor of hematology and oncology at the University of Chicago. At around the same time, Olopade combined her interests in cancer genetics, prediction, and prevention to establish the Cancer Risk Clinic, an affiliate of the University of Chicago Hospitals, which integrated cutting-edge scientific research with clinical care. In an interview for the website of the Breast Cancer Research Foundation, Olopade described the clinic's objectives as being to "coordinate preventive care and testing for healthy patients and their families who because of genetics or family history are at increased risk for cancer" and to focus "on quality-of-life issues for young breast cancer patients, including concerns related to pregnancy, fertility, and employment." The Cancer Risk Clinic's own website summed up its mission as three-fold—patient care, clinical research, and education (for medical students).

Studying the BRCA1 Gene

In the mid-1990s Olopade focused her scientific and clinical research on the area of breast cancer—the leading cause of death among American women ages forty to fifty-five—and its prevalence among young nonwhite women, particularly those of African descent. (A cousin of Olopade's had succumbed to the disease at age thirty-two.) Her efforts were motivated in part by a widely celebrated breakthrough in the greater scientific community regarding the identification of BRCA1, a gene "believed to cause about half of all inherited cases of breast cancer," as reported by Thomas H. Maugh III for the *Los Angeles Times*

(September 15, 1994). Originally pinpointed in 1990 by the geneticist Mary-Claire King of the University of California, Berkeley, BRCA1 was found to be a tumor-suppressor gene that, when healthy, prevents the aberrant replication of cells; when dysfunctional it "can certainly

> "Cancer awareness, even among physicians, and much more so among women at risk, needs an enormous boost in Nigeria."

cause cancers," as Jon Turney remarked for the *London Independent* (June 6, 1994). Citing a study published in the *Lancet*, a medical journal, Turney further noted, "A woman carrying the altered form of the gene has an 80 percent chance of contracting breast cancer before she reaches seventy." He qualified that statement, however, by adding that "only around 5 percent of all breast cancer cases are linked to mutations inherited from the patient's parents, and only around half of these are due to BRCA1. . . . Most cancer researchers think that the inheritance of a BRCA1 mutation represents one stage of a multistep process which is seen in every case of breast cancer." That qualification aside, medical professionals received news of BRCA1's discovery with great optimism. In 1995 BRCA2, a second gene linked to breast cancer susceptibility, was identified. The discovery of BRCA1 confirmed prior studies that linked heredity to cancer susceptibility and strengthened the case for genetic testing as a viable detection measure. An early and passionate advocate of genetic testing, Olopade explained in the portion of the article she wrote for the *Chronicle of Higher Education* (July 3, 1998), "It is now widely accepted that cancer is a genetic disease, and that an accumulation of genetic defects can cause normal cells to become cancerous. . . . The most effective approach to preventing, diagnosing, and treating cancer is to identify individuals who are at risk. Genetic testing for cancer susceptibility can identify these individuals and provide unique opportunities to develop new strategies for early detection and prevention."

The BRCA1 discovery introduced a host of ethical and socioeconomic questions and implications for clinical practice. With regard to genetic testing, which had become the preferred method for amassing statistics on breast-cancer susceptibility, for instance, why did the findings not reflect a more diverse sampling of patients? Olopade addressed that question in an editorial column for the *New England Journal of Medicine* (November 7, 1996): "At present, the only women at high risk who are not being tested are those who are economically disadvantaged or uninformed. The situation is clearly unfair. How should physicians responsibly address this latest challenge when there are no prospective, randomized clinical trials on which to base decisions?" In particular, Olopade cited the disparities between breast-cancer victims of African heritage and those of European ancestry, which she had observed in part through years of treating and observing patients at the Cancer Risk Clinic. While visiting Nigeria for a niece's wedding in 1997, Olopade spent time at a breast-cancer clinic and was struck by how young the patients there were, in comparison with white breast-cancer patients. From the well-documented research performed by the scientific community throughout the 1990s, Olopade noted that while African American women experienced a lower incidence of breast cancer than Caucasian women, they had developed its more virulent forms at younger ages, contributing to higher mortality rates from the disease among black women. Determined to unravel the medical mystery presented by that finding, Olopade launched a cross-continental research project that sought to collect and analyze the genetic profiles of more than 100,000 breast-cancer patients under the age of forty-five in the United States, Africa, and the Caribbean; with the resulting data she hoped to identify the environmental, social, and nutritional factors that increased a woman's susceptibility to breast cancer.

Olopade presented some of her findings in 2004 at a landmark conference she helped organize—the First International Workshop on New Trends in the Management of Breast and Cervical Cancers, held in Lagos. Drawing experts from the University of Chicago, the University of Ibadan, and the Medical Women's Association of Nigeria, the symposium addressed the issues of cancer awareness, detection, and prevention in African nations including Nigeria, whose breast-cancer patients' rate of surviving for five years stood at 10 percent. (By contrast, five-year survival rates for breast-cancer patients in the United States exceeded 85 percent.) As quoted in an online news release from the University of Chicago Hospitals (May 14, 2004), Olopade assessed the state of Nigeria's medical practices in stark terms: "Cancer awareness, even among physicians, and much more so among women at risk, needs an enormous boost in Nigeria. . . . We were trying to follow about five hundred young Nigerian women who had been diagnosed with breast cancer, but in a very short time, without access to optimal treatment, almost all of them had either died or been lost to follow-up." Cancer, Olopade observed, remained a leading cause of death for Africans, despite the overshadowing urgency of the continent's AIDS pandemic.

In 2005, over a decade after the identification of BRCA1, Olopade and other researchers from the University of Chicago produced evidence that suggested unique gene expressions in the breast cancers of African women. Through an ongoing study funded by the Breast Cancer Research Foundation and the National Women's Cancer Research Alliance, Olopade compared cancer tissue from 378 women in Nigeria and Senegal with tissue samples from 930 women in Canada, deriving a number of significant conclusions in the process. Olopade determined that breast cancers in African women, for instance, originated from basal-like cells rather than the milk-secreting luminal cells of their white American and European counterparts (tumors caused by basal-like cells are more dangerous). She further observed that most African breast cancers lacked the estrogen receptors found in 80 percent of Caucasian breast cancers. Cancer tumors in African women, as a result, were largely unaffected by hormonal stimulus and neutralized the power of such conventional, estrogen-blocking drugs as Tamoxifen. On the positive side, Olopade found that African cancers were less likely to express HER2 (or human epidermal growth factor receptor), a protein that when mutated stimulates aggressive cancer-cell growth; only 19 percent of Africans expressed that cell, compared with 23 percent of Caucasians. Those statistics have implications for African American women, 40 percent of whom "have a lineage that can be traced back to historical Nigeria," as Lovell A. Jones, a researcher at the University of Texas, told Cynthia Daniels for *Newsday* (December 30, 2005). As quoted in a University of Chicago Hospitals online news release (April 18, 2005), Olopade concluded that while breast cancer struck fewer women in Africa than in the Americas, it "hits earlier and harder." She hoped that her findings would initiate new testing measures for breast-cancer patients of varying ethnicities. "We need to reconsider how to screen for a disease that is less common but starts sooner and moves faster," she said. "Obviously, an annual mammogram beginning at fifty is not the best route to early detection in African women, who get the disease and die from it in their forties, and it also needs to be adjusted in African Americans and high-risk women of all racial/ethnic groups."

Olopade continues to refine predictive models for cancer risk that incorporate patients' genetics, ancestry, and ethnicity. "Breast cancer isn't one single disease," she told Kimberly L. Allers for *Essence* (January 2006). "It affects women of different populations in different ways." For her pioneering research in the molecular genetics of breast cancer and for the development of genetic testing measures for women of African heritage, Olopade was awarded a MacArthur Foundation grant in September 2005. On its website, the MacArthur Foundation explained the selection of Olopade, stating, "In bridging continents with her innovative research and service models, Olopade is increasing the probability of improved outcomes for millions of women of African heritage at risk for cancer here and abroad." In April 2006 Olopade was awarded the inaugural Minorities in Cancer Research Jane Cook Wright Lectureship by the American Association for Cancer Research (AACR), in recognition of her contributions as a minority physician and scientist to cancer research. At the AACR annual meeting in Washington, DC, Olopade delivered a lecture titled "Nature or Nurture in Breast Cancer Causation." According to Sola Ogundipe, writing for *Vanguard* (April 18, 2006), the lectureship honored Olopade's recent discovery of a so-called putative suppressor gene "that is involved in several other solid breast cancer tumors," a finding that "eventually led to the identification of the gene tagged p16INK4." In April 2006 Olopade traveled to China Medical University in Shenyang for a two-week visiting professorship, through which she presented her groundbreaking research on cancer genetics. Regarding her decision to lecture in China, Olopade told Bennie M.

Currie for *Crisis* (May/June 2006), "Our work on women of color has impact for other minority women, too. . . . Breast cancer is a disease that globally affects more than a million women."

Olopade is currently overseeing a trial run in West Africa for Capecitabine, a pill form of chemotherapy aimed at African women with advanced stages of breast cancer. Olopade holds memberships in a number of professional organizations, including the American Association for Cancer Research, the American Association for the Advancement of Science, and the American College of Physicians. She continues to lecture as professor of medicine and human genetics at the University of Chicago Medical Center, where she is now a full professor, and to direct the Cancer Risk Clinic at the University of Chicago Hospitals. Olopade also serves on the board of trustees of the Healthy Life for All Foundation, an advocacy group that promotes the advancement of medical technology in Nigeria.

Olufunmilayo Olopade is married to Christopher Olopade, an associate professor of medicine and director of clinical research at the University of Illinois at Chicago. They have three children and live in the Kenwood community, outside Chicago.

Further Reading

Maugh, Thomas H., II. "Ovary Removal May Cut Cancer Risk." *Los Angeles Times*. Los Angeles Times, 21 May 2002. Web. 9 Aug. 2012.

Peterson, Kristin. "Dr. Olufunmilayo Olopade to Present Commencement Speech at Dominican University." *TribLocal Oak Park/River Forest*. Tribune Company, 20 Dec. 2010. Web. 9 Aug. 2012.

Sweet, Lynn. "Another Chicagoan, Olufunmilayo Falusi Olopade, Gets Obama White House Post." *Chicago Sun-Times*. Sun-Times Media, 25 Feb. 2011. Web. 9 Aug. 2012.

Paabo, Svante

Swedish geneticist

Born: April 20, 1955; Stockholm, Sweden

"I'm driven by curiosity, by asking the questions, where do we come from, and what were the important events in our history that made us who we are," the molecular biologist Svante Paabo told Steve Olson for *Smithsonian* (October 2006). "I'm driven by exactly the same thing that makes an archaeologist go to Africa to look for the bones of our ancestors." But rather than exploring the past through the excavation of archaeological ruins or fossilized remains, Paabo uses segments of ancient DNA strands to map the key genetic changes that made humans who we are today. "Studying DNA is exciting to me, I think, because, in a way, it is like making an archaeological excavation," he said in an interview for "Family That Walks on All Fours," a November 14, 2006, installment of the PBS program *Nova*. "But you don't do it in a cave somewhere, you do it in our genes. And you find out things about our ancestors that we cannot find out any other way—that the bones and the stones will not tell us."

Considered the founder of paleogenetics—the name given to the use of genetics to study ancient populations—Paabo is making rapid strides in answering the question that has perplexed humanity for ages: what differentiates humans from other primates? "Svante is going to be the first anthropologist to win a Nobel Prize," Richard Klein, a paleoanthropologist at Stanford University in California, told Nancy Shute for *U.S. News & World Report* (January 20, 2003). "He just comes out with one paper after another that seems to be a breakthrough in human evolution." Nonetheless, Paabo has remained modest about the scope of his work. "It is a delusion to think that genomics in isolation will ever tell us what it means to be human," he told Laurent Belsie for the *Christian Science Monitor* (February 14, 2001). "The history of our genes is but one aspect of our history, and there are many other histories that are even more important."

Early Life and Education

An only child, Svante Paabo was born on April 20, 1955, in Stockholm, Sweden. His mother, Karin Paabo, was a food chemist, and his father, Sune Bergstrom, was a biochemist. Even as a child Paabo was interested in ancient civilizations, and when he was thirteen years old, his mother finally granted his most fervent wish: a trip to Egypt. "It was absolutely fascinating," Paabo told Olson. "We went to the pyramids, to Karnak and the Valley of the Kings. The soil was full of artifacts." In 1975 Paabo enrolled in Sweden's Uppsala University. He intended to study Egyptology but was disappointed by the program's emphasis on such desk work as memorizing hieroglyphic verb forms rather than performing archaeological fieldwork. "It was not at all what I wanted to do," he told Olson. Instead, Paabo switched to studying medicine and molecular immunology, and in 1981 he became a full-time PhD student in the Department of Cell Research at Uppsala.

But even as he pursued other areas of study, Paabo remained interested in his first love, Egypt. "I knew about these thousands of mummies that were around in museums, so I started to experiment with extracting DNA," he told Olson. With the help of his former Egyptology professors, Paabo obtained skin and bone samples from twenty-three mummies, and—while still a graduate student—worked in secret at night and on weekends, worried about the consequences of his actions for his academic career if he was found out. Ultimately, he extracted a short segment of DNA from a 2,400-year-old mummy of an infant boy, becoming the first scientist to isolate ancient human DNA. Paabo sent the resulting paper to *Nature*, one of the world's foremost scientific journals, which published the landmark study as its cover story.

Paabo also sent the results to Allan Wilson, a molecular biologist at the University of California, Berkeley, who was famous for first suggesting that a comparative analysis of mitochondrial DNA could be used to determine the relationships of different species to one another. Wilson, assuming that someone who had managed such an achievement as Paabo's must be a full-fledged professor, responded by asking to spend his sabbatical at Paabo's lab. Paabo wrote back, telling Wilson that he was still a graduate student and asking if he could work in Wilson's lab instead. After receiving his PhD in molecular immunology from Uppsala University, in 1986, Paabo conducted postdoctoral research at the Institute for Molecular Biology at the University of Zurich in Switzerland. He also worked briefly for the Imperial Cancer Research Fund in London, England, before joining Wilson at Berkeley in 1987, to continue his postdoctoral studies. In Wilson's lab, Paabo focused on developing new ways to extract DNA from such extinct organisms as a flightless bird called a moa and marsupial wolves. (Michael Crichton credited the group and its groundbreaking work for inspiring his bestselling 1990 novel, *Jurassic Park*.) It was at Wilson's lab that Paabo became familiar with polymerase chain reaction (PCR), a recently invented method for rapidly replicating multiple strands of a DNA sequence. PCR could turn a minuscule amount of DNA into an amount large enough to study; it turned out to be the key to allowing scientists to analyze—and not just extract—ancient DNA.

Life's Work

In 1990 Paabo was appointed to his first academic post, becoming a full professor of general biology at Ludwig Maximilian University Munich in Germany, where he continued to study the DNA of such ancient animals and plants as mammoths, European cave bears, and maize. Paabo was also part of the team that analyzed the corpse of a man who had died around 3300 BC and had been preserved in an Alpine glacier on the Italian-Austrian border. The team confirmed that "Otzi," as they named the ice man, was indeed an ancient human and not the product of a hoax. The analysis of such old samples was especially impressive given the formidable obstacles in obtaining ancient DNA. Shortly after the death of an organism, the nucleic acid that comprises its DNA begins to degrade rapidly. It is possible, however, to extract segments of DNA from a specimen that is thousands of years old, particularly when it has been preserved in stable and cold conditions in caves or in the permafrost. Even if a scientist is able to locate a well-preserved specimen, he or she must contend with the problem of contamination. The modern humans who handle specimens leave traces of their own DNA, which is extremely difficult to discern from that of their ancient counterparts.

As the analysis of ancient DNA grew more commonplace, Paabo gained a reputation as not only one of the technique's pioneers but also one of its most reliable practitioners and stalwart defenders. At a conference on ancient DNA at the University of Oxford in England, Paabo urged accountability for testing results, according to Nigel Williams for *Science* (August 18, 1995): "The field has established itself and shown that you can go back in time and gather data on mo-

> **"I'm driven by curiosity, by asking the questions, where do we come from, and what were the important events in our history that made us who we are."**

lecular evolution," Paabo said. But, he warned his colleagues, "if the field is to become fully respectable it is vital to ensure the reproducibility of results."

In the mid-1990s Paabo began making significant strides in what would become his greatest project to date, mapping Neanderthal (*Homo neanderthalensis*) DNA. Neanderthals—human ancestors whose remains, dating from roughly 30,000 to 200,000 years ago, have been found in Europe, northern Africa, and western Asia—coexisted with modern humans (*Homo sapiens*) for tens of thousands of years. Scientists have long debated what role Neanderthals played in the evolution of modern humans. Most anthropologists currently support the out-of-Africa theory, which contends that modern humans originated in Africa, spreading out over the globe and gradually forcing Neanderthals into extinction. Another camp, made of up those who support the multiregional theory of human development, proposes that Neanderthals and modern humans interbred, which would mean that Neanderthals did not die out but rather were subsumed into the modern human species.

Paabo's own research into that debate provided significant evidence to bolster the out-of-Africa position, when, in 1997 (some sources state 1996), a team of American and German scientists that he led became the first to extract mitochondrial DNA from Neanderthal remains.

(Mitochondrial DNA is more abundant and consequently easier to extract than nuclear DNA. For example, the typical human cell contains 500 to 1,000 copies of mitochondrial DNA and only two copies of nuclear DNA. Mitochondrial DNA, however, contains DNA only from the mother's side.) Using a bone fragment from the Neanderthal specimen that was found in Neandertal, Germany, in 1856, Paabo and his team were able to extract and analyze one sequence of Neanderthal DNA. That one sequence differed drastically from the same stretch of DNA in a modern human, suggesting that Neanderthals and modern humans were in fact two separate species that had diverged approximately 600,000 years ago (some sources state 500,000 or 550,000) with little or no interbreeding. "It's a fantastic achievement," Christopher Stringer, a leading proponent of the out-of-Africa theory, told Curt Suplee for the Austin (Texas) *American-Statesman* (July 11, 1997). "In terms of our knowledge of human origins, it's as big as the Mars landing." Even the most forceful supporters of the multiregionalism theory took notice, influenced by Paabo's impeccable research record. Milford Wolpoff, a harsh critic of the out-of-Africa theory, was quoted by Suplee as calling Paabo's results "exciting." Wolpoff continued, "If anybody could do this beyond criticism, it's Svante Paabo and his laboratory."

While continuing his work on Neanderthal DNA, Paabo has also been involved in a number of other important projects. In 2001 Paabo and his colleagues published a study that was intended as a response to Edward Hooper's hypothesis—detailed in *The River: A Journey to the Source of HIV and AIDS* (1999)—that researchers had accidentally sparked the AIDS epidemic by using HIV-infected chimp kidneys to manufacture experimental polio vaccines that were distributed in Africa during the 1950s. Paabo and his fellow researchers tested samples of the polio vaccine for traces of the HIV virus or chimpanzee DNA, finding neither. "I feel confident that we have done everything we can to detect chimp DNA in this sample, so as far as I can see there is not much more one can presently do in order to clarify the situation further," Paabo told Raja Mishra for the Boston *Globe* (April 26, 2001).

Paabo's work, however, continued to focus primarily on discerning what separates humans from other primates. Describing his research, he told *Current Biography*, "We work on understanding various aspects of what makes humans unique from a genetic perspective, for example by studying gene expression in various tissues in humans and apes." Apes—and specifically chimpanzees—constitute the best point of comparison because they are our closest living relations. Modern humans and chimpanzees have evolved separately since they split from a common ancestor between five and seven million years ago. Still, the two species have 98.7 percent of their genes in common. "I'm still sort of taken aback by how similar humans and chimps are," Paabo told Malcolm Ritter for the Associated Press (August 31, 2005). "I'm still amazed, when I see how special humans are and how we have taken over this planet, that we don't find stronger evidence for a huge difference in our genomes." Human and chimpanzee genomes may differ slightly, but, according to a paper by Paabo published in *Science* in 2002, many of the genes that the species share operate differently in human brains from the way they function in chimpanzee brains. Paabo's findings potentially explain why the two species have such different mental capacities. Though Paabo insisted that the results were not absolutely conclusive, they did single out the brain as a specific focal point for human evolution.

Max Planck Institute for Evolutionary Anthropology

In 1997 Paabo and three other collaborators were invited to start a new institute on human evolution, the Leipzig-based Max Planck Institute for Evolutionary Anthropology, which was generously funded by the German government. Paabo's codirectors were a linguist, a comparative psychologist, and a specialist in wild chimpanzees, and the combination of specialties complemented the interdisciplinary nature of his own pursuits. Inspired by a 2001 finding that linked a mutation in the gene FOXP2 to human speaking disorders, Paabo began comparing the gene in humans and other primates. His team's findings, published in *Nature* (August 22, 2002), confirmed that the FOXP2 gene carries mutations unique to modern humans. Those differences became widely established among humans within about the last 200,000 years, and while not directly responsible for the emergence of speech, the gene probably allowed our ancestors to speak more clearly. "We had communication of a sort, and then this change happened and allowed the carriers to articulate much better," Paabo explained to Alex Dominguez for the Associated Press (August 14, 2002). "This may have been the time we arrived at truly modern, articulate language." That ability to communicate, Paabo told Dominguez, may have been the key to the survival and eventual dominance of modern humans, giving those who carried changes in the gene an advantage in hunting, for example. Still, not all the mysteries of the FOXP2 gene were resolved, and Paabo rejected the hypothesis that the gene alone allows for speech. "People might say if we put this in a chimpanzee, it could talk," he told Dominguez. "I don't think that is the case, speech is more complex than that."

In 2006 Paabo and his team announced that, using a novel method, they had extracted the first sample of nuclear DNA from Neanderthal remains. Later that year, in November, they published their findings—an analysis of one million units of Neanderthal DNA—in *Nature*. Paabo's team released its findings simultaneously with another team, led by Edward Rubin of the Lawrence Berkeley National Laboratory in California, which had independently analyzed 62,250 units of DNA from Paabo's sample and published the results in *Science*. Accounting for acceptable margins of error, the two teams agreed that Neanderthals and modern humans share 99.5 percent of their DNA, making Neanderthals modern man's closest genetic relation. While the 0.5 percent discrepancy seemed to further bolster the out-of-Africa position, neither team was willing to entirely dismiss the possibility that humans and Neanderthals interbred—at least until more of the Neanderthal genome can be sequenced. Paabo announced plans to have a draft of the full genome—a total of 3.2 billion units—ready in two years. "The next big step is the retrieval of sequences of entire genomes of extinct creatures," he told *Current Biography*. "And that revolution is about to happen as we speak."

A foreign member of the United States' National Academy of Sciences and other academies, Paabo has received many awards and honors, including the Prix Louis-Jeantet de Medecine, the Ernst Schering Prize, the Max Delbruck Medal, and the Newcomb-Cleveland Prize. Paabo was listed among *Time* magazine's "100 most influential people in the world" in 2007. In 2010, he received the Theodor Bücher Medal from the Federation of European Biochemical Societies.

Paabo, who was described by Michael Dumiak in *Archaeology* (November/December 2006) as "a bristly browed Swede with a penchant for wearing sandals and goofy socks," lives in Leipzig, Germany, with his partner, Linda Vigilant, and the couple's son, Rune Vigilant.

Further Reading

Kolbert, Elizabeth. "Sleeping With the Enemy." *The New Yorker.* Condé Nast, 15 Aug. 2011. Web. 9 Aug. 2012.

Olson, Steve "Neanderthal Man." *Smithsonian.com.* Smithsonian Media, Oct. 2006. Web. 9 Aug. 2012.

Shute, Nancy. "The Human Factor." *USNews.com.* U.S. News & World Report, 20 Jan. 2003. Web. 9 Aug. 2012.

Rubin, Edward M.

American geneticist

Born: February 17, 1951; New York, New York

James Watson, the famed scientist who codiscovered the double-helix structure of DNA, predicted that the international effort to map the human genome "will change mankind, like the printing press," according to a reporter for the BBC (June 26, 2000). Now that an international team of scientists has carefully transcribed the three billion base pairs that comprise the human genome—the "Book of Life," as scientists call it—geneticists such as Edward M. Rubin, the director of the US Department of Energy's Joint Genome Institute and the genomics division at the Lawrence Berkeley National Laboratory, are left with the daunting task of finding the meaning in the text.

Rubin is a pioneer in the study of DNA sequences that do not code for genes but rather play an important role in regulating gene expression. "I've been very interested in the part of the genome that does not code for proteins," he told *Current Biography*. "Some call this junk DNA, but clearly it's not all junk and some is functional. About 98 percent of the genome might be categorized as junk DNA and we really don't know how to read meaning in that part of the genome." Through the use of large-scale cross-species DNA sequence comparisons, he has indeed proven that there are "jewels of function amongst the junk," as he says. "The hotbed of medical research was the gene itself," Rubin told Sophia Kazmi for the Contra Costa (California) *Times* (October 17, 2003). "Now there is reason to look outside."

Early Life and Education

Known to associates and friends as Eddy, Edward M. Rubin was born in New York City on February 17, 1951. He and his older sister, Penny, were raised on the outskirts of the city. His father, Sol, was an accountant who also ran a small business, and his mother, Irene, was a homemaker. Even as a child Rubin took an interest in science, working on science projects and playing with his home chemistry set. He received a bachelor's degree in physics from the University of California, San Diego in 1974. After reading James Watson's *Molecular Biology of the Gene*, he was inspired to focus his studies on genetics. "[The book] made me appreciate for the first time how beautiful the machinery of life was with DNA, RNA and the building of proteins eventually leading to cells and organisms. Simple and elegant, in some ways like an erector set," Rubin told *Current Biography*. "Suddenly, in a flash, I realized that biological systems did not have infinite complexity, but were decipherable." In 1980 he earned an MD and a PhD in biophysics from the University of Rochester School of Medicine, in New York State. He then served a clinical residency and a genetics fellowship at UC-San Francisco. Later, he worked as a research associate at the Howard Hughes Medical Institute in Chevy Chase, Maryland.

Life's Work

In 1988 Rubin started working at the Lawrence Berkeley National Laboratory, an arm of the US Department of Energy that is managed by the University of California. Two years after Rubin's arrival at the lab, the Energy Department and the National Institutes of Health launched the historic Human Genome Project, a collaborative effort by hundreds of researchers at colleges, universities, and laboratories throughout the world to catalog, or "sequence," all of the three billion chemical base pairs that make up human DNA. The project was completed ahead of schedule, in 2003—fifty years to the month after James Watson and Francis Crick discovered the double-helix shape of DNA; the Berkeley Lab had contributed significantly to the project, sequencing 11 percent of the human genome (more specifically, chromosomes 5, 16, and 19).

Rubin, who played a major role in the final phase of sequencing the human genome, has worked throughout his scientific career to develop computational and biological approaches to decipher the information encoded by the sequence of the genome. To that end, in 1992 he founded the Rubin Lab at the Berkeley National Laboratory. There, using transgenic mice—mice that have had DNA from another species, usually *Homo sapiens*, inserted into their own DNA—Rubin and his staff attempted to identify the genetic links to such common ailments as heart disease. "You can't look at heart disease in a test tube, so these transgenic animals have a very major impact," he told Marla Cone for the Los Angeles *Times* (May 9, 1993). "They contribute to our understanding of how things work in a setting fairly close to the way it works in people. It's really been a gold mine for researchers." On May 1, 1997, in the journal *Nature Genetics*, Rubin and his research associates reported their discovery of a major genetic factor contributing to mental retardation in Down syndrome, which affects an estimated one million Americans and is the leading cause of mental retardation in the United States. Most individuals afflicted with Down syndrome carry a complete extra copy of chromosome 21, which also results in problems with the immune and endocrine systems, as well as skeletal and tissue deformities, but because there are rare forms of the syndrome in which patients carry only a portion of chromosome 21 in triplicate, Rubin suspected that only a limited number of genes on the chromosome were responsible for mental retardation. To test his hypothesis Rubin's team inserted different adjacent segments of human chromosome 21 into mice and then tested their learning skills. The researchers found that inserting an additional copy of a particular gene, DYRK, affects the ways in which neural pathways are constructed and impairs learning ability. "Our strategy made no prior assumptions about individual genes, but rather, allowed the behavior of the mice to guide us to a crucial gene on chromosome 21," Rubin told a reporter for *Gene Therapy Weekly* (June 2, 1997). "This approach of using the whole animal is likely to become increasingly important as geneticists attempt to investigate the basis of other complex human conditions such as hypertension or schizophrenia."

Later in 1997 Rubin and Tim Townes of the University of Alabama at Birmingham reported in *Science* (October 31, 1997) that their labs had found a way to replicate sickle-cell anemia in mice. Every year thousands of human babies are born with sickle-cell anemia, a sometimes fatal disease in which a mutant form of hemoglobin distorts red blood cells into a sickle (crescent) shape, which makes it difficult or impossible for the cells to pass through small blood vessels. In time, without the normal flow of blood, such tissue becomes damaged.

Victims of the disease suffer intense pain. Although scientists have known for more than fifty years which mutant gene causes the disorder, they have yet to develop a cure. Several scientists had attempted without success to introduce the mutant gene into mice in the 1980s and early 1990s, in the hope of being able to test a greater range of possible treatments than would be permitted on human subjects. Franklin Bunn, an authority on hemoglobin diseases of the Harvard Medical School in Boston, Massachusetts, told Marcia Baringa for *Science* that Rubin and Towne's breakthrough was "a critically important advance" in the search for a treatment for sickle-cell anemia.

In 1998 Rubin was appointed head of the Genome Sciences Department at UC-Berkeley. The following year his research team reported in *Nature Genetics* (October 1, 1999) that by using a new technique, they had isolated two genes that increase an individual's odds of contracting asthma. Instead of relying on the usual "brute-force method" of comparing the DNA sequences of individuals afflicted with asthma with those from people who are not, the team examined sets of genes on chromosome 5 associated with allergic activity, inserting them into mouse DNA. Each mouse carried a different portion of the chromosome, allowing Rubin and his team to identify genes IL4 and IL13 as those responsible for asthma susceptibility by simply observing the outcomes in the mice.

Understanding "Junk" DNA

Before scientists had set about sequencing the DNA of humans and animals, it was thought that genes—segments of DNA that build, or "code," proteins (the building blocks of life)—were the crucial elements of the genome. That raised a puzzling question, because 95 percent of the human genome consists of noncoding DNA, otherwise known as "junk" DNA. Some scientists hypothesized that those DNA "deserts," as these long stretches of junk DNA were called, had once served an evolutionary purpose but were now dormant. As scientists began sequencing the genomes of mice, rats, and humans, they discovered matching patterns in the DNA deserts of all three species. The idea emerged that if those segments had been preserved long after mice and rats diverged from the same evolutionary chain, and humans diverged from rodents, then those segments must present some evolutionary advantage. Scientists then discovered that some segments of noncoding DNA could be used to regulate the expression of particular genes. In *Science* (April 7, 2000), Rubin and his associates at the Berkeley Lab and UC-San Francisco reported that the technique Rubin's lab had been using to identify genetic links to disease could also be used to mine DNA deserts for regulatory DNA. "You could call this finding jewels in junk DNA," Rubin told a reporter for *AScribe Newswire* (April 11, 2000). "By comparing human and mouse sequences we can identify those segments of the genome that contain information which instructs surrounding genes on when and where they are to be active. Identifying these regulatory sequences using classical biological approaches is labor intensive and difficult." In a subsequent issue of *Science* (October 18, 2001), Rubin and his team announced that by using that method they had discovered such a "jewel"—an apolipoprotein that plays an important role in regulating triglyceride levels in the blood. Triglycerides and cholesterol, the two major blood fats, pose important risk factors in the development of heart disease, the leading cause of death in the United States.

The next year the Rubin Lab, in conjunction with twenty-seven other institutions in six countries, completed a project that sequenced the mouse genome and compared it with the human genome. Published in *Nature*, the study revealed that mice share a genetic heritage with humans; depending on various measurements, the number of common genes might be as high as 99 percent of the genome and could be traced back 75 million years. The mouse was the second mammal to be decoded, and its genetic similarity to humans provided scientists with the opportunity to further investigate DNA deserts; the research team was surprised to discover that almost three times as much of the genome common to mice and men lay in DNA deserts as in genes. "Now instead of a landscape with no common points, we have a map that scientists can travel over," Rubin told Ian Hoffman for the Oakland (California) *Tribune* (December 5, 2005). "Now they have really vital highways into all kinds of biomedical research."

In January 2003 Rubin was appointed director of the US Department of Energy's Joint Genome Institute and the genomics division at the Lawrence Berkeley National Laboratory. The Walnut Creek, California-based Joint Genome Institute (JGI) employs about 240 people and is managed by the Berkeley Lab in conjunction with two other UC-affiliated facilities: the Lawrence Livermore National Laboratory,

> "The ability to compare DNA sequences in the human genome [with] sequences in nonhuman primates will enable us in some ways to better understand ourselves than [will] the study of evolutionarily far-distant relatives such as the mouse or the rat."

near San Francisco, California, and the Los Alamos National Laboratory, in New Mexico. Rubin had been serving as the interim director since the spring of 2002 and was selected for the position after a nationwide search.

Phylogenetic Shadowing

On February 28, 2003, in *Science*, the JGI and the Berkeley Lab announced the development of what is known as phylogenetic shadowing, a new technique for deciphering the biological information encoded in the human genome. The technique enables scientists to make meaningful comparisons between the DNA of humans and monkeys, apes, or other primates. Previously, contrasting the human genome with those of chimpanzees or baboons had been difficult, because the sequences are so much alike—only 2 percent of the chimp sequence is different, and only 5 percent of the baboon's—and it is difficult to distinguish which segments encode functional elements. Instead of comparing the genetic sequence of one primate with that of the human genome, Rubin and his team compared segments from the genomes of several different primates and found enough small differences among species to compile a phylogenetic shadow. Using the shadow, Rubin and his team were able to identify DNA sequences that regulate the activation of a gene that is found only in primates and is an important indicator of an individual's risk of developing heart disease. "The ability to compare DNA sequences in the human genome [with] sequences in nonhuman primates will enable us in some ways to better understand ourselves than [will] the study of evolutionarily far-distant relatives such as the mouse or the rat," Rubin told a reporter for *Genomics & Genetics Weekly* (March 28, 2003). "This is important because as valuable as models like the mouse have been, there are many physical and biochemical attributes of humans that only other primates share."

Later that year, in *Science* (October 17, 2003), researchers at the Berkeley and JGI labs led by Rubin announced that the sequences in DNA deserts that regulate particular genes and the genes being regulated may be located surprisingly far from one another in the genome. "The distance from which these long-range enhancers can reach out across the genome to regulate a gene is a hundredfold greater than anyone thought," Rubin told a reporter for *Genomics & Genetics Weekly* (November 7, 2003). "Gene deserts may not be home to any genes but they can host DNA sequences that act as long-distance switches to activate far-away genes. This suggests that the idea that all gene deserts could be eliminated with no consequences to the organism is wrong." Rubin's team reached this conclusion after studying a human gene called DACHI, which is involved in the development of the brain, limbs, and sensory organs. Because DACHI resides between two large gene deserts, Rubin thought that regulatory elements for the DACHI gene might lie in the adjacent deserts. By comparing the human and mouse genomes, the researchers were able to identify seven long-range gene enhancers buried in the surrounding gene deserts. The discovery may prove useful in the treatment of diseases. "Regulatory elements that are very far away from a gene can cause serious problems," Rubin told the reporter for *Genomics & Genetics Weekly*. "You can think of these long-range enhancers like the root system of a tree. Previously we thought that roots only extended a short distance from the tree. Our new study suggests that some roots can extend far away from the tree's trunk and branches, [and that] cutting those distant roots can still do harm to the tree."

In a study published in *Nature* in October 2004, however, Rubin demonstrated that certain portions of the DNA desert could be deleted without any negative effects. Rubin's team deleted large segments of seemingly functionless stretches of mouse DNA, producing mice that exhibited no apparent abnormalities. The 1,234 noncoding sequences deleted from the mice were identical to sequences found in the human genome. "In these studies, we were looking particularly for sequences that might not be essential," Rubin told a reporter for the Wellcome Trust's website (October 20, 2004). "Nonetheless we were surprised, given the magnitude of the information being deleted from the genome, by the complete lack of impact noted. From our results, it would seem that some noncoding sequences may indeed have minimal if any function." In similar studies, parts of the genome of similar sizes had been deleted in mice, but because those sequences contained important genetic information, none of those mice survived. "If you're an architect and you want to see if a wall is weight bearing, you remove the wall and see if the ceiling caves in," he told Bryan Nelson for *Newsday* (October 21, 2004). "That's what we did and were expecting the roof to fall in, and it didn't." The discovery could also prove useful to scientists scouring the human genome for genes that cause or contribute to disease: Rubin's results could help them to eliminate large stretches of the three billion base-pair human genome that are likely irrelevant. "Looking for genes involved in human diseases is like looking for a needle in a haystack," he told Betsy Mason for the Contra Costa *Times* (October 21, 2004). "Here's a big part of that haystack where you don't have to look."

Rubin served as the lead researcher for a study in which his team demonstrated that the signatures of genes in terrestrial and aquatic environmental samples could be used to diagnose the health of environments. As described in *Science* (April 22, 2005), those DNA fingerprints, called Environmental Genomic Tags, or EGTS, serve as a DNA profile of a particular site and reflect the presence of levels of nutrients, pollutants, and other environmental features. For example, microbial creatures living in surface water rely heavily on light as a source of energy; consequently, samples taken from shallow seawater contain an abundance of genes involved in photosynthesis, the process by which sunlight is converted into energy. "EGT fingerprints may be able to offer fundamental insights into the factors impacting on various environments," Rubin said, as quoted in a JGI press release (April 21, 2005). "With EGTs we don't actually need a complete genome's worth of data to understand the functions required of the organisms living in a particular setting. Rather, the genes present and their abundances in the EGT data reflect the demands of the setting and, accordingly, can tell us about what's happening in an environment without knowing the identities of the microbes living there." Instead of sequencing the genomes for all of the microbes in the environment—a daunting task, given the large number of different organisms present in nearly all environments—the EGT approach allows scientists, by means of an abbreviated sequencing technique, to form a useful metabolic picture of an environment. "Environmental systems are extremely complex, harboring numerous diverse species coexisting in a single locale," Rubin explained in the press release. "By focusing on the information encoded in the DNA fragments sequenced, independent of the organisms from which they derived, we were able to get around the problem of species diversity."

On June 2, 2005, in an online supplement to *Science*, Rubin and other Berkeley Lab researchers announced a breakthrough that could help resolve the controversy over the relationship between modern humans and Neanderthals (*Homo neanderthalensis*). Working with James Noonan of the Berkeley Lab and experts in ancient DNA in Austria and Germany, Rubin sequenced portions of nuclear DNA taken from two specimens of an extinct cave bear (*Urus spelaeus*) that died in Austria more than 40,000 years ago. Previously, scientists had been unable to reconstruct the genome of an extinct animal, because the nucleic acid that comprises DNA begins to degrade rapidly after death; moreover, such samples are easily contaminated with the DNA of the people who find it and the microbes that devoured the animal after it died. In rare instances, in which specimens had been unusually well preserved in permafrost or snow, scientists had succeeded in extracting nuclear DNA from their remains, but only from specimens less than 20,000 years old.

Many of the previously successful studies of ancient DNA had focused on mitochondrial DNA, which is more abundant and easier to extract than nuclear DNA. For example, the typical human cell contains 11,000 copies of mitochondrial DNA and only two copies of nuclear DNA. Mitochondrial DNA, however, is much less useful to scientists; it contains only DNA from the mother's side, and scientists need information from the father's DNA, too, to understand biological differences between species. "While mitochondria are great for learning about evolutionary relationships between species, to understand the functional differences between extinct and modern species we really need nuclear or genomic DNA," Rubin told David Gilbert for US Fed News Service (June 3, 2005).

Rubin and his team were able to reconstruct the cave bear genome using a tooth from one animal and bone from the other. They sequenced all of the DNA in the two samples, including the contamination. (It turned out that only 6 percent of the DNA sequenced from the sample contained the cave bear's DNA sequence; the other 94 percent came from various contaminants.) Aided by computers, the researchers compared the cave bear DNA with that of a dog—which diverged from bears around 50 million years ago but still share 92 percent of their genes with bears—so as to pinpoint what is identical to the cavebear sequences in the lengthy sequences of human and microbial DNA. "We were looking for the proverbial needle in the haystack . . .," Rubin told Lee Bowman for the Scripps Howard News Service (June 2, 2005). "Among the expected lion's share of contaminants we recovered reasonable amounts of 40,000-year-old cave bear DNA and useful information from it. We were lucky in that we had a very powerful magnet in the form of industrial strength computing to tease out the interesting data from a hodgepodge of different DNAs."

The sequence produced was only about 1/1,000th of 1 percent of the total bear DNA, but according to the team, it proved that given sufficient time and resources, the researchers could indeed sequence the genome of an entire extinct animal. "This is very much a proof in principle," Rubin told a staff writer for *New Scientist* (June 11, 2005). "We're not interested in cave bears—we're interested in Neanderthals." For the past century there has been a fierce debate surrounding the exact relation of modern humans to Neanderthals, hominids that lived in Europe and Asia during roughly the same period as cave bears. There is some evidence that Neanderthals, humans' closest prehistoric relatives, coexisted with early people. Though they possessed shorter, broader bodies than ours, their anatomy was so similar that in 1964 it was suggested that modern humans and Neanderthals actually represent subspecies: *Homo sapiens sapiens* and *Homo sapiens neanderthalensis*. This view held great currency in the 1970s and 1980s, but many anthropologists today have returned to the two-species hypothesis.

It will be more difficult but, the team believes, not impossible to distinguish Neanderthal DNA from human DNA that has contaminated a Neanderthal sample. On July 6, 2005, the Max Planck Institute for Evolutionary Anthropology announced the launch of a joint project between German and US scientists to reconstruct the Neanderthal genome. Rubin, who was also participating in the project, spoke of the effort to the German weekly *Die Zeit*, according to a reporter for the Associated Press Worldstream (July 6, 2005). "Firstly, we will learn a lot about the Neanderthals," he said. "Secondly, we will learn a lot about the uniqueness of human beings. And thirdly, it's simply cool."

Rubin has served as cochair of the Cold Springs Genome Sequencing and Biology Meeting and was scientific chair of the International Human Genome Organization. He lives in Berkeley with his wife, Joan Emery, a genetic counselor. They have two teenage children, Ben and Rachel. In his leisure time Rubin likes to surf in the nearby Pacific Ocean.

Further Reading

"JGI Director Eddy Rubin." *DOE Joint Genome Institute.* The Regents of the University of California. Web. 9 Aug. 2012.

"Neanderthal Genome Sequencing Yields Surprising Results and Opens a New Door to Future Studies." *Research News.* Berkeley Lab, 15 Nov. 2006. Web. 9 Aug. 2012.

Swaminathan, Nikhil. "Genomic 'Time Machine' May Pinpoint Divergence of Human and Neandertal." *Scientific American.* Nature America, 15 Nov. 2006. Web. 9 Aug. 2012.

Solomon, Susan

American atmospheric scientist

Born: January 19, 1956; Chicago, Illinois

"As a young student of chemistry in the late 1970s, I became intrigued with the notion that chemistry could be explored on a planet instead of in a test tube," the atmospheric scientist Susan Solomon wrote for the website of the University Corporation for Atmospheric Research. Solomon embraced that idea in two journeys to Antarctica—in 1986 and 1987—that were critical in providing evidence of damage to the ozone layer, which has since been widely acknowledged as one of the most daunting threats to the environment. For her work in revealing the effects of industrial chemicals on the ozone, Solomon was awarded the National Medal of Science, the highest honor presented to scientists by the United States government. Solomon has also been active in the study of global warming, the gradual heating of the atmosphere caused by man-made pollutants. In addition, she is the author of *The Coldest March*, a nonfiction book about the ill-fated journey to Antarctica by the explorer Robert Falcon Scott.

Early Life and Education

Susan Solomon was born in January 1956 in Chicago, Illinois. Her father was an insurance salesman; her mother was a fourth-grade schoolteacher. Solomon has felt an inclination toward the sciences "since I was about ten years old and I watched [the undersea explorer] Jacques Cousteau on TV," as she told Julie Hutchinson for the Chicago *Tribune* (May 24, 1992). She excelled in chemistry in high school and majored in that subject at the Illinois Institute of Technology in Chicago. She graduated with a bachelor's degree in 1977 and spent that summer working at the National Center for Atmospheric Research (NCAR). She received her PhD in chemistry from the University of California, Berkeley in 1981. That same year she began working as a research chemist for the US government's National Oceanic and Atmospheric

Administration (NOAA), where she has spent more than twenty years investigating the causes of the depletion of the ozone layer.

Hole in the Ozone

Often referred to as a "natural sunscreen," the ozone layer lies in Earth's stratosphere (between 10 and 50 kilometers above the surface). Ozone (O_3) forms when ultraviolet radiation from the sun splits oxygen molecules (O_2) into atomic oxygen (single oxygen atoms, O); the atomic oxygen combines with O_2 to form O_3. The ozone layer blocks certain components of the sun's rays, such as ultraviolet rays, which can cause skin cancer in humans. In 1973 Sherwood Rowland and Mario Molina, chemists at the University of California, Irvine, began to test the influence of chlorofluorocarbons (CFCs), industrial chemicals used in refrigerators, air conditioners, and aerosol spray cans, on the stratosphere. Over time they formed the theory that CFCs might be causing the ozone layer to thin, allowing more of the sun's ultraviolet rays to pass through the atmosphere. Rowland immediately spoke out about his discovery, calling for a ban on all CFCs; he was met largely with skepticism and ridicule, in part because he and Molina lacked hard evidence of what they suspected. In 1984 Anne Burford, the head of the Environmental Protection Agency during President Ronald Reagan's administration, called the claims of ozone depletion a "scare issue"; several years later that administration's secretary of the interior, Donald Hodel, suggested that any danger presented by a hole in the ozone layer was minimal, and that people could protect themselves simply by wearing sunglasses and sunscreen. In 1985, however, a team of British scientists discovered that the ozone layer over Antarctica had been depleted by about 30 percent in ten years, which was far faster than anyone had thought, and that the hole in the layer was deepening each spring.

In 1986 Solomon, greatly interested in the findings of Rowland and Molina, took center stage on the issue. She presented a new theory, which was that while chlorofluorocarbons could be harmful to ozone, it was the intense cold of the Antarctic that was accelerating the process to such an extent. Since the deepening of the hole occurred in the spring, as British scientists discovered and Solomon's team confirmed, the researchers theorized that the damage was done in the late winter; studying this phenomenon meant heading into Antarctica, the planet's coldest region, during its coldest season. Solomon thus led a research team into the Antarctic in the heart of winter in an attempt to record data on how CFCs affect the ozone layer.

In an interview with Gabrielle Walker for *New Scientist* (October 13, 2001), Solomon explained that she had been excited to visit the cold, desolate region: "I loved it from the minute I stepped off the airplane," she said. "I loved the wildness of it. I loved the pristineness, the emptiness, the ferocity, the absolutely unforgiving nature of that environment. I love the things about nature that remind us how powerful it is and how puny we are, and Antarctica is as fierce and beautiful as you can get." Her work there proved to be difficult. "This is one of the most challenging things that we've ever come across in atmospheric chemistry," Solomon said to James Gleick for the *New York Times* (July 29, 1986) just before her trip. "Whatever the source [of the ozone depletion] is, we need to understand it because this is a change in the ozone that's of absolutely unprecedented proportions. We've just never seen anything like what we're experiencing in the Antarctic."

Solomon studied the way in which the Antarctic air absorbed moonlight during the long winter nights. To do so she set up a system of mirrors to reflect moonlight—and exposed herself to extremely harsh conditions. "The mirrors were on the roof of our laboratory about a mile from the main base," Solomon told Gabrielle Walker. "One night I was observing and about 3 a.m. it clouded over so I went to sleep. When I woke up in the morning a howling blizzard was under way." Solomon headed out into the intense winds to save the mirrors and continue the research. "I went up there and tried to get the things down, and the wind was gusting up and I was pushed several feet across the roof. I dropped down spreadeagled and crawled back and finally got the mirrors down."

The expedition, which extended from August to November 1986, was met with considerable interest from the media. Michael Parfit wrote for the Los Angeles *Times* (August 31, 1986), "The voices that come floating up from the bottom of the Earth will be using words like ozone and chlorofluorocarbons and stratosphere, but they will be talking about human survival." The experiments conducted by Solomon's team did not lead them to immediate conclusions about the causes of the ozone damage, and the scientists announced that it would be necessary to take another trip the following year. By the end of her 1987 trip, Solomon was more certain of her findings. "The evidence right now," she said in an interview for the CBS News program *West 57th*, as quoted by John Horn in the Orange County *Register* (July 4, 1987), "strongly suggests that [chlorofluorocarbons] are the most likely cause of the ozone hole."

In addition to supporting that belief, Solomon's experiments bolstered the theory that the ozone hole appeared over Antarctica because the intense cold accelerated the chemical reactions that took place due to the CFCs. She also asserted that the eruption of the volcano Mount Pinatubo, in the Philippines, in 1991 had contributed to the depletion of ozone over Antarctica. "Volcanic particles make chlorine from CFCs more effective at ozone destruction," Solomon said to Mark Steene for the *Advertiser* (October 1, 1998).

In 1987, fueled by rising evidence that CFCs were responsible for the life-threatening hole in the ozone layer, the United Nations Environment Program sponsored a gathering in Canada of forty-six nations whose industries produced significant amounts of CFCs. At the meetings, collectively called the Montreal Protocol, the countries' representatives agreed to freeze CFC use at existing levels; the accords were revised and expanded in 1992, calling for progressive cuts in CFC use. Meanwhile, the ozone hole continued expanding; in 1998 it was measured at three times the size of Australia. Its growth was attributed largely to the prior buildup of CFCs, which take about fifty years to dissolve. "It's going to be several decades before we see the end of the Antarctic ozone hole," Solomon told Mark Steene. She expressed some optimism, however, noting that the decrease in CFC emissions would eventually lead to the closing of the hole, and adding, "If you had to have an ozone hole," the area above the uninhabited Antarctica was "probably the best place to have it."

For her role in identifying and legitimizing the notion of ozone depletion and identifying its causes, Solomon has received many accolades and awards. In 1994 a glacier in Antarctica was christened the Solomon Glacier, in her honor. In March 2000 she was awarded the United States' highest scientific honor when then-president Bill Clinton presented her with the National Medal of Science. Upon being named for the award, Solomon reacted with modesty. "I was in the right place in the right time with the kind of science opportunity that

> "I love the things about nature that remind us how powerful it is and how puny we are, and Antarctica is as fierce and beautiful as you can get."

only happens very rarely, a once-in-a-scientific-lifetime," she said, as quoted by Ann Schrader in the Denver *Post* (February 1, 2000). William Daley, then secretary of commerce, called Solomon "one of the most important and influential researchers in atmospheric science during the past fifteen years," as quoted by the Environment News Service (February 4, 2000). In June 2004, Solomon received the Blue Planet prize, a prestigious award (which includes $460,000) presented by the Asahi Glass Foundation of Japan to scientists dedicated to preserving the environment.

The Coldest March

In 2001 Solomon published *The Coldest March: Scott's Fatal Antarctic Expedition*, a book about the explorer Robert Falcon Scott, who in 1912 was in a race with Roald Amundsen to lead the first team to reach the South Pole. Amundsen beat Scott by a month; on Scott's return trip, he and his team of four men froze to death. Scott described the expedition in great detail in his diary, which was found on his body. In the years after his death, Scott was portrayed by later explorers and authors as an incompetent man whose mistakes were responsible for the tragedy. Solomon probed more deeply into the story of Scott, having come across his cabin in Antarctica during one of her trips there. After reading his diaries, she decided to look further into the atmospheric conditions that existed in Antarctica in the year he died, and her findings inspired her to write the book. Solomon sought to provide a more balanced view of Scott, claiming in the book that he and his team had encountered conditions no one could have survived. "No one could have predicted the persistent cold weather that Scott faced," she explained to Gabrielle Walker. "In seventeen years of direct data from the area where he died, there's only been one year like that. George Simpson, Scott's meteorologist, was convinced that the weather was highly unusual, but he couldn't prove it and I could. That's when I knew I had to write this book." Solomon explained that during Scott's expedition, "the average daily low should have been around –30 degrees Celsius in March. That feels toasty once you've acclimatized. But –40 degrees Celsius is an entirely different matter. You feel as if your nostrils are being freeze-dried, and frost forms at the edge of your eyes. That kind of cold has really serious consequences for an expedition." In a review of the book for the *Spectator* (October 13, 2001), M. R. D. Foot wrote, "Over and over again, Solomon picks on the legends that have accumulated round Scott to his disadvantage, and disproves them." By the end of the book, according to Foot, Solomon has vindicated Scott and his team to the extent that "we can truly believe they were heroes."

Work at the International Panel on Climate Change

In 2002 Solomon was selected to be a part of the Intergovernmental Panel on Climate Change (IPCC), an international group established by the World Meteorological Organization (WMO) and the United Nations Environment Program (UNEP) to examine changes in the atmosphere over time. Working with the IPCC, Solomon turned her attention toward another potentially fearsome phenomenon: global warming. Often confused with theories about the hole in the ozone layer, the global-warming theory holds that the burning of fossil fuels and other human-generated pollution are responsible for a slow and gradual warming of Earth's atmosphere. The issue of global warming is politically charged, and Solomon has been careful to provide balanced assessments of the dangers the phenomenon may pose. (In a controversial move, the United States refused to adopt the Kyoto Accords, through which more than 160 nations agreed to try to curb the production of greenhouse gases. In March 2001 President George W. Bush declared that lowering emissions of carbon dioxide as called for in the agreement would be too costly for American companies.) As quoted on the IPCC website, Solomon described the need for scientists to address the issue, saying, "I . . . don't think we want to go down in history as the generation that thought it might be causing the climate to start to change but just couldn't be bothered to try and figure out whether that was true or not." She said, as quoted by the Associated Press (June 23, 2004), "The serious debate that is ongoing now is how much global warming we will have if we continue to put CO_2 [carbon dioxide] into the atmosphere." On the other hand, she has also rejected claims that certain environmental disasters, such as a lethal heat wave in France in 2003, were due to global warming. "People who attributed France's heat wave last year to global warming have been left with egg on their faces," she said, as quoted by Colin Patterson in the Wellington, New Zealand, *Dominion Post* (September 4, 2004). Above all, she has stressed that with regard to such issues as global warming, science should not be mixed with politics. "I don't think we should detract from the dignity of our science by making political statements," Solomon told *Current Biography*. "As scientists we have to be objective and focused, not political." In 2007, the IPCC was awarded the Nobel Peace Prize in recognition of "their efforts to build up and disseminate greater knowledge about man-made climate change."

Solomon lives in Boulder, Colorado, where she works for NOAA and continues to research atmospheric phenomena. She is married and has one stepson.

Further Reading

Pachauri, Rajendra. "The 2008 TIME 100: Susan Solomon." *TIME Magazine.* Time, 30 Apr. 2009. Web. 10 Aug. 2012.

Revkin, Andrew C. "Melding Science and Diplomacy to Run a Global Climate Review." *New York Times.* New York Times Company, 6 Feb. 2007. Web. 9 Aug. 2012.

Walker, Gabrielle. "In From the Cold." *New Scientist* 13 Oct. 2001: 4646. Print.

Stefansson, Kari

Icelandic geneticist

Born: 1949; Reykjavik, Iceland

Dr. Kari Stefansson, an Icelandic neuropathologist, left behind a tenured position at Harvard University Medical School in Boston, Massachusetts, to pursue the genetic origins of disease with his own company. Unhindered by the financial limitations of a university laboratory, the company, deCODE Genetics Inc., uses Iceland's comprehensive genealogical records, combined with the country's medical records and donated DNA samples, to pinpoint the genes that cause hereditary diseases. Stefansson's methods have led to a chorus of protests from privacy advocates and medical ethicists around the world, many of whom are suspicious of the Icelandic government's motives in granting deCODE access to the nation's medical records and the right to sell the resulting statistical data. Prior to Stefansson's return, Iceland's only major industries were fishing and aluminum smelting, and his biotechnology firm has brought jobs and money along with it. Stefansson's opponents thus contend that financial incentive may have forced the current agreement, which they believe represents a threat to privacy and sets a precedent for greater human-rights transgressions in the future. Despite such controversy, Stefansson's project represents an important milestone in genetic research.

Early Life and Education

Kari Stefansson was born in Reykjavik, Iceland, in 1949. His father was a popular journalist, author, and member of the Althing, Iceland's parliament. Today, Iceland has one of the highest standards of living in the world, but when Stefansson was young, its economy was so weak that apples could be imported only for Christmas. "For a thousand years we were desperately poor," Stefansson told Nicholas Wade for the *New York Times* (June 18, 2002). "And in spite of that we were convinced we had a great culture. This nation lived by that, by the

[Norse] sagas. That may be one reason genealogy became so important."

In 1976 Stefansson earned his medical degree from the University of Iceland and then moved to the United States, where he did his specialization in neurology and neuropathology at the University of Chicago. (Neuropathology is the study of diseases of the nervous system.) In 1983 he became a professor of neurology, neuropathology, and neuroscience at the University of Chicago. He completed a postdoctoral thesis in 1986, earning an additional degree from the University of Iceland. He remained on the staff of the University of Chicago until 1993, when he left to take a position as professor of neurology, neuropathology, and neuroscience at Harvard University Medical School. He also took over as director of neuropathology at Boston's Beth Israel Hospital.

Life's Work

At Harvard, Stefansson worked to discover the complex genetic causes of multiple sclerosis, a disease that creates lesions on the myelin sheath protecting nerve fibers in the spinal cord. The deterioration results in numbness, paralysis, and sometimes death. Stefansson had the idea of studying the genetic roots of the disease—not within a family as had previously been done—but within a large population, over many generations. He applied to the National Institutes of Health (NIH) for a grant to pursue this line of research back in his native Iceland, where conditions would be fortuitous for such a search because of the homogeneity of the population; the application was denied. At the time, geneticists believed that only studies of siblings and other close relatives would be statistically relevant to the hunt for disease-causing genes. Stefansson grew frustrated. "The classical university labs are simply too small to do advanced research anymore," he told Mary Williams Walsh for the *Los Angeles Times* (June 5, 1998). "The only way I saw it financially possible to continue would be to establish a company."

Founding deCODE Genetics

In 1995 Stefansson sought investors for a proposed biotechnology startup and eventually gathered $12 million from several American venture capitalists. In 1996 he founded deCODE Genetics in

Reykjavik, Iceland. (He left Harvard soon after.) Initially, the company hired twenty employees, with Stefansson as chief executive officer and president. "I have no regrets about leaving Harvard to do this," Stefansson told Mary Walsh. "We are adding new knowledge.... We are building a new industry in Iceland."

In the mid-1990s biotechnology researchers, often funded by large pharmaceutical companies, began studying remote, homogeneous populations with a high incidence of particular diseases. They hoped that disease-causing genes would be easier to detect against a uniform genetic background with little ethnic variety. The approach had some early successes; the gene responsible for Huntington's disease, which causes memory loss and abnormal body movements, was found by researchers studying an isolated community in Venezuela. Stefansson believed that Iceland's unique situation made it an even more valuable genetic resource than other populations. Vikings from Norway, along with a bounty of captured Celtic women from Ireland, settled this remote, treeless, volcano-studded island in the North Atlantic in 874 CE, and there have been relatively few new settlers since. Three disasters—a bubonic plague outbreak in 1402, a smallpox epidemic in 1708, and a volcanic eruption in 1784—greatly diminished the Icelandic genetic pool in three successive sweeps. "If you go back to the fifteenth century, you can more or less connect everybody to the same ancestors," Hakon Gudbjartsson, the chief computer engineer at deCODE, told Mary Walsh. Iceland's population of approximately 275,000 has another advantage over similar communities: traditionally, Icelanders have been very interested in genealogy, and extensive databases go back thousands of years. In addition, medical records for every citizen have been routinely archived since the country's socialized healthcare system debuted in 1915. Since 1930, tissue samples have been stored from every autopsied body. Stefansson realized that, in Iceland, he would be able to track disease-causing genes across wide swaths of extended families over long periods of time. To do this, he would need to create a "phenotype database" that would compile all of the information. Powerful computers could then search for patterns in a process called "data mining." "You may have someone with diabetes in your grandfather's generation," Stefansson explained to a reporter for *Technology Review* (April 1, 2001), "and it skips the parents and then it affects the children, and so on. And to study these disorders, you have to go to a population, because once a disease begins to skip generations, a nuclear family isn't a useful unit of society to study." If the Icelandic people, one of the most technologically savvy populations in the world, could be convinced to cooperate, Stefansson knew that his ambitious plans could become a reality. "We have an opportunity to do the best human genetics done anywhere in the world," Stefansson told Walsh.

There was a great deal of controversy in scientific and medical circles about giving access to the nation's medical and genealogical records to a private company seeking profits. Some critics called Stefansson's idea "bio-piracy," accusing him of being exploitive. In his defense Stefansson pointed out that the company's access to Icelandic medical records was similar to that given to any researcher doing an epidemiological study. (Epidemiology is the statistical examination of disease in a population.) "When you go to your doctor to seek medical help," Stefansson explained to Ira Flatow for the National Public Radio show *Talk of the Nation* (January 19, 2001), "you are drawing on knowledge and experience that was gathered because your parents and their parents allowed information like this to be used with presumed consent. If they had not done so, if informed consent had been a prerequisite for the use of information like this, you would not have a health-care system of the type that we have today." Critics weren't satisfied and argued that if medical records were stolen or released somehow, health-insurance companies might then be tempted to deny coverage to people with a history of disease. (This claim, repeated frequently in the press, was spurious because of Iceland's socialized health care. It would only be a concern if deCODE's methods were adopted elsewhere.)

The company's method of gathering data was designed for both efficiency and privacy. First, patient identities are encrypted by the Icelandic Ministry of Justice's Privacy and Data Protection Authority (PDPA), which allows deCODE to match information between databases with anonymous identification numbers instead of names. Through relationships with Icelandic doctors, the company assembles the medical records of patients with a disease being currently studied. Using the encrypted identity codes, researchers are then able to cross-reference these records with an extensive genealogy database, compiled by deCODE from a variety of sources, in order to place the patients into the context of the Icelandic family tree. When researchers find a number of individuals with the same disease in a hereditary cluster—sometimes a group of people who don't even realize they are related through a distant ancestor—they contact the patients' doctors. The doctors, in turn, ask their patients to participate in the deCODE study. In this way, no one at the company learns the identities of patients. If the patients agree to participate, doctors obtain DNA samples from them and disease-free relatives and send them off to deCODE to be statistically compared. The method proved itself soon after deCODE's founding, when researchers determined the main gene responsible for familial essential tremor, a disease that causes uncontrollable shaking in 5 to 10 percent of all people over the age of sixty-five. Surprisingly, the search took only three months. Researchers using traditional techniques had failed to accomplish this goal after years of trying.

By January 1998, deCODE had hired one hundred employees. Icelandic scientists began returning from prestigious posts all over the world to work at the firm. In February of that year, Icelandic Prime Minister David Oddson passed the pen between Stefansson and Jonathan Knowles of Hoffman-LaRoche as they signed a deal worth an estimated $200 million to deCODE. Hoffman-LaRoche, the Swiss pharmaceutical giant, would help fund the Icelandic firm's research in return for the rights to develop new drugs based on its findings. Some saw Oddson's ceremonial position at the signing as an indication of overly enthusiastic government support for deCODE. Their suspicions only deepened when the Althing introduced a piece of legislation, called "A Bill on Medical Databases," that would grant an unnamed company, widely understood to be deCODE, the right to create and market a database containing every Icelander's medical records. (Although the bill was created specifically for deCODE, Icelandic law requires that government contracts be open to competitive bidding. One other company bid for the right to create the database, but deCODE won.) Computer terminals would be installed in every medical office in the country, at which doctors would upload disease and prescription information to a central database, encrypted by the PDPA for future cross-referencing with genealogy and DNA samples. This would greatly accelerate deCODE's process, because without a central database each doctor had to be contacted individually for information. The company would then have monopoly rights to the information in the database for twelve years. In return, it would pay the Icelandic government $12 million a year. Alfred Arnason, the

> **"Here we are, a tiny little biotech company on a rock in the North Atlantic, and we have put together a map that can vastly improve the human genome."**

chief geneticist at a Reykjavik tissue-typing lab, objected to the bill, which he believed boiled down to selling medical information that had previously been available free to researchers. "That would hinder science, not enhance it," Arnason told Mary Walsh. "If he wants to chase genes, that's all right. Everybody is chasing genes nowadays. But I'm in favor of everybody having the same right to do their studies and not monopolize data." (Technically, deCODE would not be able to monopolize the data; rather, it would be able to license and sell its own database of compiled information. The original data would be available to other researchers.) Arnason's was just one voice in the uproar; an estimated seven hundred articles about the bill appeared in Iceland's three newspapers over the course of the debate. *Mannvernd*, a group of scientists and physicians opposed to the bill, sent an open letter to the government urging them to delay any deal with deCODE until the issues were considered more carefully. The initial bill was blocked, but only minor changes were made before the Althing passed a revised bill in December 1998. (The caveat was added that Icelanders could choose to remove themselves from the database. By 2001, however, only seven percent of the population had done so.)

In 1999 deCODE officials announced promising results with both osteoarthritis and pre-eclampsia, a condition that causes critically high blood pressure during pregnancy. Critics complained that these successes weren't nearly as important as the company was claiming. "In order to develop useful drugs from such information," Matt Crenson wrote for the Associated Press (March 21, 2001), "deCODE would need to pinpoint the genes themselves, figure out what they do, and develop ways of affecting that activity. Then they would have to develop the drugs to do the job and demonstrate their safety and effectiveness to the appropriate regulatory agencies, a process that can take decades." Stefansson dismissed his scientific critics. "These are people who wish they had thought of these ideas themselves," he told Kitta Macpherson for the *Hamilton Spectator* (December 29, 2000). The company's board of directors showed their enthusiasm for Stefansson's progress by electing him chairman, in addition to his roles as president and CEO. Despite skepticism in the scientific community, enthusiasm for deCODE continued to grow among Icelanders and Icelandic businesses. The company was not registered on any official stock market yet, but individual investors and businesses were able to buy shares in deCODE on Iceland's "gray" stock market, which exists thanks to a loophole in the country's securities law. By March 2000 the share price had grown in value thirteen-fold. Nearly every Icelander had a stake in the future of deCODE, whether they had invested directly or through their pension funds.

In 2000, the company announced its intention to go public on the NASDAQ stock exchange. Shortly thereafter, US president Bill Clinton and British prime minister Tony Blair jointly called for free public access to raw genetic information. This announcement was a contributing factor to a sudden downshift in biotechnology gains. Stefansson almost immediately announced the discovery of a gene connected to stroke, but there was little he could do to stem the industry's rapid decline. The company's stock price sank from $65 to $18 by the time it went public in July 2000. Despite this drop, deCODE managed to raise $244 million with its initial public offering. A poll of Icelanders at the time showed that overall support for deCODE's work had risen from 75 to 91 percent.

By 2001 deCODE had grown to employ four hundred people. Company scientists had discovered genes responsible for Alzheimer's disease, osteoporosis, and peripheral arterial occlusive disease. In addition, Stefansson began licensing deCODE's software tools for encryption and gene tracking to other biotechnology firms as a way to generate short-term revenue. Applied Biosystems, a manufacturer of DNA sequencing machines, adopted deCODE's genotyping software, and several pharmaceutical companies now pay for a deCODE service that maps genetic reactions to drugs. Stefansonn's methods have been so successful that the company has been able to correct large errors in the rough map of the human genome released by the Human Genome Project in 2000. "Here we are, a tiny little biotech company on a rock in the North Atlantic, and we have put together a map that can vastly improve the human genome," Stefansson told Nicholas Wade.

By 2002 almost 80,000 people (a substantial percentage of the population) had participated in one or more deCODE studies. Hoffman-LaRoche extended its contract to 2006, expanding its scope beyond gene discovery to drug development for four separate diseases. In July 2002, after a three-year search, deCODE scientists located a gene connected to schizophrenia, a disease that affects Stefansson's brother. The company acquired two American laboratories—one on Bainbridge Island in Washington State, and one in Chicago, Illinois—to handle drug development and medicinal chemistry in its search for cures. Eventually, Stefansson plans to ask all Icelanders to voluntarily donate blood for DNA sampling, instead of approaching people individually based on their disease profile. "Actually that would be the least invasive method to obtain DNA," he explained to the *Technology Review* reporter. "When you're doing it through physicians who take care of patients and relatives of patients, there is a certain coercion involved because you're approaching people at a moment of at least perceived need."

Today, deCODE's population-based data-mining methods have been adopted by several other biotechnology firms, including Wellcome Trust of London, which announced plans in 2002 to create a database to study disease in the English population. Still, many scientists and doctors continue to call the company's phenotype database a threat to human rights. Stefansson firmly defends his methods. "I have visited about twenty-five countries this year, and in every country I come to, I take out my ATM card and withdraw money," he told the *Technology Review* reporter. "I can do that because the entire world is a network of centralized databases of personal information on finance." Stefansson sees a centralized medical database as far less threatening. "You actually have to have a fairly lively imagination to figure out how you can abuse it [in healthcare]. . . . The only kooks that could abuse this would be kooks in the insurance industry. We can simply regulate that."

Currently deCODE has targeted fourteen specific genes that cause about a dozen diseases. In October 2003, Hoffman-LaRoche and deCODE announced the discovery of a gene mutation that can double the risk of heart attack. The following month deCODE released the news of three gene variants that can increase the risk for osteoporosis by 30 percent. Drug development is now being pursued on peripheral arterial occlusive disease, stroke, and schizophrenia. A cure for obesity is being researched in partnership with Merck, another pharmaceutical giant; a milestone in deCODE's obesity research was announced in September 2003: geneticists had discovered a gene that, in one form, predisposes a person to obesity but in another form to thinness. In 2007, Stefansson was included in the "Time 100," *Time* magazine's annual list of the one hundred most influential people in the world. In 2009, Stefansson received the European Society of Human Genetics Award. The same year, Stefansson also received the Andres Jahre Award for Medical Research, one of the most prestigious medical prizes in Europe.

Stefansson lives in Reykjavik with his wife and three children. "My family has lived in Iceland for 1,100 years," he told the *Technology Review* reporter. "There is a certain adaptation that has taken place. We fit this sort of wet, barren, dark corner of the North Atlantic." In his spare time, he is an avid reader of poetry.

Further Reading

Caplan, Jeremy. "The TIME 100: Kari Stefansson." *Time.* Time, 3 May 2007. Web. 9 Aug. 2012.

Sample, Ian. "Rare Genetic Mutation Gives Hope In Battle Against Alzheimer's." *The Guardian.* Guardian News and Media, 11 July 2012. Web. 9 Aug. 2012.

Wade, Nicholas. "Gene May Play a Role in Schizophrenia." *New York Times.* New York Times Company, 13 Dec. 2002. Web. 9 Aug. 2012.

Walsh, Mary Williams. "A Big Fish in a Small Gene Pool." *Los Angeles Times* 5 June 1998: A1. Print.

Steitz, Joan A.

American biochemist

Born: January 26, 1941; Minneapolis, Minnesota

"A personal romance"—that is how Joan A. Steitz has described her relationship with ribonucleic acid (RNA), a component of cells that plays a crucial role in the construction of proteins, which are required for the execution of all the basic biological processes of the human body and from which muscle, nerves, blood vessels, bone, and everything else in the body is made. Steitz, a biochemist, has been studying RNA since the early 1960s, and her work has remained at the forefront of RNA research for many years. She is best known for discovering in the cell nucleus particles identified as small nuclear ribonucleoproteins—snRNPs, pronounced "snurps," consisting of RNA plus a protein—and for illuminating the function of snRNPs in the synthesis of a form of RNA known as messenger RNA, or mRNA. In particular, she discovered that snRNPs handle a step in that synthesis called splicing. Steitz's work has huge ramifications for medical science, not least because defects in the splicing process may account for 10 to 15 percent of all genetic diseases. Colleagues of Steitz's as well as other interested observers have predicted that her discoveries will lead to new ways of diagnosing and treating autoimmune diseases (in which the immune system, which is designed to protect an organism against disease and infection, mistakenly attacks the organism itself) and other disorders.

Steitz holds the title Sterling Professor of Molecular Biophysics and Biochemistry at the Yale University School of Medicine—Yale's highest academic rank—and heads a laboratory at the Howard Hughes Medical Institute (HHMI) at Yale's Boyer Center for Molecular Medicine. John Dirks, a professor of medicine at the University of Toronto and the president of the Gairdner Foundation, told Elaine Carey for the Toronto (Ontario) *Star* (April 3, 2006) that Steitz is "one of the most distinguished molecular biologists of her time and has been an outstanding role model to scientists and students everywhere." She is also a "tireless promoter of women in science," as the biochemist Christine Guthrie described her in an interview for the *ASCB Newsletter* (June 2006), and she "has made it a priority" to give up-and-coming female scientists "something she didn't have: a female network within the system," as Margaret N. Woodbury wrote for the *HHMI Bulletin* (February 2006). In 2006, in recognition of her extraordinary contributions to biomedical science, Steitz won a Gairdner Foundation International Award and the National Cancer Institute's Rosalind E. Franklin Award for Women in Science—two of the several dozen honors she has received. Among the others are the Eli Lilly Award in Biological Chemistry (1976), the National Medal of Science (1986), the Weizmann Women in Science Award (1994), the UNESCO-L'Oreal Award for Women in Science (2001), the New York Academy Medal for Distinguished Contributions in Biomedical Science (2008), and twelve honorary doctorates.

Early Life and Education

The daughter of Glenn D. and Elaine (Magnusson) Argetsinger, both schoolteachers, Steitz was born Joan Elaine Argetsinger on January 26, 1941, in Minneapolis, Minnesota. She showed an early inclination toward science at a time when very few women had made careers in that field, and her teachers and family encouraged her to pursue that interest. After her graduation from a girls-only high school, Steitz attended Antioch College in Yellow Springs, Ohio, where her chemistry classmates included the future evolutionary biologist and essayist Stephen Jay Gould and the future biochemist and textbook author Judith G. Voet. Antioch's acclaimed science program included three-month "co-ops" each year, in which students worked in research labs around the country. During one of her co-ops, Steitz worked in the laboratory of Alex Rich at the Massachusetts Institute of Technology (MIT), where she was introduced to the emerging science of molecular biology. A second co-op brought her to the Max Planck Institute, in Germany. Although Steitz's fascination with science increased during her undergraduate years, she was unsure about committing herself to the extremely long hours and weekends that laboratory research invariably demanded. Furthermore, the career prospects for a female researcher seemed virtually nonexistent; she herself knew no female senior scientists or full professors. "I decided that my best future course was to attend medical school. I had known several women physicians, so this seemed a reasonable choice," Steitz wrote for the website Agora: For Women in Science (April 6 , 2006). Steitz gained acceptance to Harvard Medical School in Boston, Massachusetts, and planned to begin her studies there in the fall of 1963, a few months after she earned a bachelor's. degree.

During the summer of 1963, Steitz engaged in research in the University of Minnesota laboratory of the embryologist Joseph Gall, who gave her a project to work on—not with other people, as had always happened in the past, but independently. Her subject was *Tetrahymena pyriformis*, a one-celled protozoan that has two nuclei and is covered with hundreds of cilia (hairlike structures used in feeding, swimming, and other activities) and that, like the fruit fly, has been the focus of hundreds of scientific investigations. Gall asked Steitz to determine whether *Tetrahymena*'s cilia contain nucleic acid—a task that relied on the recent finding that mitochondria possess their own complement of DNA. (Mitochondria are among the organelles of cells. The other organelles in cells—each of which has a specific function and may be considered somewhat analogous to organs of a body—include the nucleus, ribosomes, the nucleolus, Golgi apparati, and centrioles.) Steitz told Elaine Carey, "All of a sudden I got completely turned on. I decided even if I wasn't ever going to be able to do this like all the men professors I'd known, that this was what I wanted to do, pursue science. It really, really got me and I couldn't believe how much fun it was making discoveries." Recognizing her aptitude, Gall urged Steitz to transfer from Harvard's medical school to the university's new graduate program in biochemistry and molecular biology (BMB). One faculty member in the program was James D. Watson, a friend of Gall's, who, along with Francis Crick and Maurice Wilkins, had won the 1962 Nobel Prize for Physiology or Medicine for their discovery of the molecular structure of DNA. With Watson's help, Steitz was admitted into the program. "As [Steitz has] become more and more famous, it's been gratifying to think that I had a small part in influencing her," Gall told the *ASCB* interviewer. "She just had that summer in my lab. Still, my guess is that she would have gone into research, no matter what."

In graduate school Steitz had no female professors; she was the only woman in her BMB class of ten at Harvard and the first female graduate student in Watson's lab. Under Watson's direction Steitz began to study RNA structure and function in bacteriophages (viruses, also known as phages, that are without nuclei and that infect bacteria). Seeking an adviser who would oversee her doctoral research,

> **"We cannot expect to capture the interest and talents of girls and women for the scientific enterprise unless they can view their own participation as possible."**

she approached a "famous, well-respected, and now deceased" male scientist, as she described the man—whom she declined to name—to Margaret Woodbury. The scientist said he would not work with her, on the grounds that she was a woman. Steitz recalled to Woodbury that she ran from the room and burst into tears, but that afterward she came to regard the experience as extraordinarily fortunate: Watson became her PhD adviser, and their professional relationship strengthened into a lifetime bond. "He was an excellent mentor and very supportive of my work," Steitz said of Watson in an undated interview posted on the HHMI website. "He was truly an inspiration and taught me to focus on the important questions in science." For her PhD thesis, Steitz examined in vitro (that is, in test tubes) a bacteriophage known as R17 and shed light on how protein and nucleic-acid components of viruses come together. While at Harvard she met Thomas A. Steitz, a microbiologist and X-ray crystallographer; the two married in 1966. In 1967 Joan Steitz earned her PhD.

Life's Work

That year the Steitzes moved to Cambridge, England, where Joan Steitz, as a National Science Foundation and later Jane Coffin Childs Memorial Fund postdoctoral fellow, joined Francis Crick and the molecular biologist Sydney Brenner (winner of the Nobel Prize in 2002) at the Medical Research Council Laboratory of Molecular Biology. There, working with bacteriophages, Steitz investigated the exact point (the so-called start point) on a strand of messenger RNA on which a ribosome binds and the manufacture of protein begins. Messenger RNA is one of several types of RNA synthesized within the cell in the process known as transcription, which is guided by the genetic information, or codes, in DNA. In her first independent discovery, Steitz located three start points on bacteriophage mRNA. As the sole author, she described her discovery in "Polypetide Chain Initiation: Nucleotide Sequences of the Three Ribosomal Binding Sites in Bacteriophage R17 RNA," in *Nature* (December 6, 1969).

Back in the United States in 1970, Steitz embarked on a lecture tour and was surprised to find that many universities were interested in hiring her. A profusion of job offers came her way in part because of the attention that her *Nature* paper had attracted. Another cause of the interest in her was that in recent years, American universities had begun to add increasing numbers of women to their math and science faculties, in response to growing pressure from within and outside academia. That same year Steitz accepted a position as an assistant professor at Yale University in New Haven, Connecticut; her husband also joined the Yale faculty. "I was exceedingly nervous about accepting a 'real job' because I knew I had not prepared myself to be a faculty member in the same way that my male colleagues had," Steitz wrote in an article for the Agora: For Women in Science website (April 6, 2006). At Yale, Steitz continued her work on bacteriophage RNA binding sites before shifting her focus to eukaryotic cells (which house genetic information within a membrane-bound nucleus). Her paper "How Ribosomes Select Initiator Regions in mRNA: Base Pair Formation Between the 3' Terminus of 16S rRNA and the mRNA During Initiation of Protein Synthesis in *Escherichia coli*," cowritten by Karen Jakes and published in the *Proceedings of the National Academy of Sciences* (December 1, 1975), explains how ribosomes identify the start site on a strand of mRNA. Steitz regards that discovery as a highlight of her career.

DNA Splicing and snRNPs

The work for which Steitz is most famous, however, was performed later in the 1970s and is described in another seminal paper: "Are sn-RNPs Involved in Splicing?" published in *Nature* (January 10, 1980), which she cowrote with four others. That work made use of the discovery (also in the late 1970s, by scientists at MIT and Cold Spring Harbor Laboratory in New York, among other facilities) that the DNA in eukaryotic cells contains combinations of two types of sequences, known as exons and introns, that alternate within a gene. Exons code for the synthesis of proteins; that is, they contain the information necessary for its construction. Introns do not code for proteins; indeed, they are referred to as "nonsense segments" of DNA or "junk" DNA. (Steitz later found, however, that on occasion an intron includes a small nucleolar RNA found in a snRNP. Those molecules chemically modify ribosomal RNA and are essential to its function.) The process of creating proteins starts with the synthesis of what is called pre-mRNA, which contains both introns and exons. Messenger RNA, created from pre-mRNA in a later phase of the process, contains only exons. Nobody yet knew the mechanism by which the introns are removed from the pre-mRNA. Steitz discovered, first, that antibodies from patients suffering from the autoimmune disease lupus reacted with previously unknown particles—the snRNPs. That observation led to her hypothesis that snRNPS, operating within the nucleus, are part of what has been named the spliceosome—the complex of molecules that removes the introns from pre-mRNA and splices together the remaining exons to form messenger RNA. "We did the first experiments that showed they were, in fact, involved in splicing," as Steitz told Woodbury. "[Steitz's] insight was a true inspiration," according to Susan Berget of the Baylor College of Medicine, as quoted in the *ASCB Newsletter*. "Her hypothesis set the field ahead by light years and heralded the avalanche of small RNAs that have since been discovered to play a role in multiple steps in RNA biosynthesis." Steitz is currently studying viral snRNPS as well as other effects of splicing on RNA in its activities later in the process of protein synthesis. She and her colleagues are also investigating types of snRNPS that are involved in excising a rare class of introns. Medical scientists have translated her findings into clinical uses that Steitz regards as, in her words, "absolutely amazing." One team, for example, found a way to use aberrant splicing to prevent the effects of muscular dystrophy in dogs. "Basically, they designed a snRNP to undo the drastic consequences of a mutation. I think that is just extremely cool," Steitz told Woodbury.

Steitz directs the work of two dozen people in her lab, among them seven graduate and undergraduate students and, in the summer, high-school students. "Part of the reason Yale is so great is that it has such fabulous undergraduates. Getting them into your lab is such a joy," she told the *ASCB* interviewer. Steitz makes sure that, in addition to group investigations, each person conducts his or her own project, "even if it's very simple, to get them hooked on how much fun it is," as she told Carey. "It's very valuable for the future and for getting women into science," she noted. At Yale, and particularly since she became a department head, she has seen firsthand the many ways in which scientists' careers can be affected by subjective decisions on the parts of their colleagues and others, regarding matters ranging from the question of who deserves extra laboratory space, a personal secretary, or research assistants, to whom should be granted a trip to a major conference, where presenting a paper confers at least some degree of prominence and offers the possibility of making useful contacts, to who merits tenure. Steitz found that women were rarely the beneficiaries of such decisions. "Unless you really know what's going on inside the department, it all looks perfectly reasonable," Steitz told Kate Zernike for the *New York Times* (April 8, 2001). "If a woman is a star there aren't that many problems. If she is as good as the rest of the men, it's really pretty awful. A woman is expected to be twice as good for half [the pay]." Steitz wrote for the Agora website, "Even today, I find from conversations with young women that they are frightened by the prospects of what lies ahead because so few women seem to have 'made it' in academic science without encountering difficulties [because of their gender]. We cannot expect to capture the interest and talents of girls and women for the scientific enterprise unless they can view their own participation as possible. . . . Our challenge now is to devise more effective practices in our universities and other research venues for capturing and advancing women in the scientific hierarchy."

Woodbury described Steitz as having "a particular smile that flickers beneath her rosy, yet elegant cheekbones when [she is] talking about or with her students. It's something her students notice and appreciate. The smile plays there as a message of encouragement, endorsing their right to think aloud even as they sometimes fumble with their biological formulations." Steitz told Woodbury, "Almost every time I lecture at another university, someone comes up to me and says, 'I took your biochemistry course back in 19xx, and it was terrific.' What more can one wish for?" Steitz's husband is a Sterling Professor of Molecular Biophysics and Biochemistry and professor of chemistry at Yale. Steitz told an HHMI interviewer that having a spouse who can empathize with "the pressures and joys of a career in academic science makes it all possible." The couple has one son, Jon, who played minor-league baseball for a few years after earning a bachelor's degree from Yale and who is currently attending Yale Law School.

Further Reading

Carey, Elaine. "Female Scientist A Hero in Her Field: Yale's Joan Steitz, 65 Honoured." *Toronto Star* 3 Apr. 2006: A4. Print.

"Joan Argetsinger Steitz." *ASCB Newsletter* Jun 2006: 18–20. Print.

Woodbury, Margaret A. "Trailblazer Turned Superstar." *Howard Hughes Medical Institute Bulletin* 19.1 (Feb. 2006): 21. Print.

Tanaka, Koichi

Japanese chemist

Born: August 3, 1959; Toyama, Japan

On October 9, 2002, the Royal Swedish Academy of Sciences announced that it had awarded the 2002 Nobel Prize in Chemistry to Koichi Tanaka, John B. Fenn, and Kurt Wuthrich, for their improvements to the methods of detection and study of large, complex molecules such as proteins. For Tanaka, who had been working in relative obscurity with the Shimadzu Corporation for many years, the news came as a great surprise. "I got a phone call in English and I could make out 'Nobel something' and 'congratulations,' but I really didn't understand," Tanaka was quoted as saying by Doug Struck and Sachiko Sakamaki in the *Washington Post* (October 12, 2002). Tanaka then rushed to a press conference, where he apologized for meeting with the press in his work clothing.

In winning the Nobel Prize at age forty-three, Tanaka became the youngest chemistry laureate since 1934. Following the award, he became something of a media darling, charming reporters with his unfeigned surprise and joy at winning the unexpected prize. He was subsequently appointed general manager of the Shimadzu Corporation's mass-spectrometry research laboratory. When he received his award, Tanaka was reportedly most excited to meet Jimmy Carter, the winner of the Nobel Peace Prize in 2002, whom he greatly admires.

Early Life and Education

The chemist Koichi Tanaka was born on August 3, 1959, in Toyama, Japan. Tanaka's biological mother died a month after his birth, from complications stemming from delivering a child at her advanced age. Because his biological father was also older and unable to tend an infant alone, his aunt and uncle raised him. Both his aunt and uncle, whom he called his mother and father, kept the knowledge of his birth parents from him until he was eighteen years old. In an autobiographical statement posted on the official Nobel Prize website, he recalled, "my blissful ignorance was due to the completely fair-minded upbringing I received by my parents [aunt and uncle], older sister and brothers, relatives, and neighbors." Tanaka's aunt and uncle, who ran a business selling and repairing carpentry tools in Toyama, instilled in him a belief in hard work, dedication, and perseverance in the achievement of one's goals. He was also greatly influenced by his grandmother, who taught him to appreciate the value of things. ("Mottai-nai," she often told him whenever he was acting wastefully, a Japanese expression that translates as "what a waste.") Growing up in Toyama, an area of Japan surrounded by beautiful mountain vistas, helped Tanaka acquire a lasting appreciation for the natural world. He explained in his autobiography, "Just the sight of the Tateyama Mountains brings me a sense of calm. This area is blessed with bountiful nature, eliciting in me a feeling of awe towards all living things and a compelling interest in the mysteries of nature, so difficult to find in urban settings."

In 1966, Tanaka was enrolled in the Hachininmachi Elementary School in Toyama City. By his own admission he was not a particularly devoted student; his interest in science first blossomed after visiting Japan's first world fair, Expo '70, in Osaka. In his autobiographical statement he recounted, "The Exposition site displayed the future of technology, which would actually be realized twenty to thirty years later. I truly felt the power of science and technology." After Tanaka began attending Toyama Municipal Shibazono Middle School in 1972, he dedicated himself to his schoolwork. By graduating in the top 10 percent of his class, he was able to enroll in the Toyama Chubu High School, a prestigious institution whose graduates typically entered top-ranking universities. He excelled as a high school student, passed his university entrance exams, and was able to gain entrance into Tohoku University, his first choice.

Upon Tanaka's acceptance to Tohoku University in 1978, his aunt and uncle needed to submit an official copy of his birth certificate to the university. They chose that moment to reveal to him that he was not their biological son. He recalled in his autobiography, "The truth came as a considerable shock to me, and the trauma surged over me in waves for a long time afterwards. At the same time, however, it was a chance for me to assert my independence. The thought grew strong in me that since I had gone to the trouble of being born, I might as well be useful in helping people live long and healthy lives. And this thought has always resided in the back of my mind."

Because he had an interest in electricity, which stemmed from his youthful hobby of building radios, Tanaka entered Tohoku University's Department of Electrical Engineering. As Japan was then in the midst of an electronics-driven business boom, he believed that a degree in electrical engineering would be invaluable in finding a job upon graduation. After receiving his degree in 1983, he began working in the Central Research Laboratory at the Shimadzu Corporation, a manufacturer of analytical instruments based in Kyoto. He was assigned to the laboratory for electrical systems-related research and joined a team of researchers responsible for developing component technology for analytical instruments.

Life's Work

Throughout the 1980s Tanaka explored ways to enhance the study of mass spectrometry—a technique that chemists have long used to identify the contents of samples by determining the mass of the molecules of which they are composed. The technique consists of ionizing (imparting an electrical charge to) a molecule and then using an electric field to separate the ions through a drift space and into a detector. Smaller ions reach the detector first, whereas larger ions take longer. By observing the time the ion takes to reach the detector, its mass and, thus, its identity can be ascertained.

When Tanaka began his work on mass spectrometry, scientists were unable to use the technique to identify most proteins, which are largely responsible for the basic biological functions of all living organisms, including bacteria, plants, and animals. Mass spectrometry could be used to identify only molecules with a weight of up to several thousand Daltons, because small, simple molecules are relatively easily ionized in preparation for analysis. (The mass of a carbon atom 12C is defined as 12 Daltons.) Larger, more complex molecules (known as macromolecules), such as proteins—which typically weigh more than 10,000 Daltons—are easily decomposed by these processes.

Tanaka's objective was to find a way to turn macromolecular samples into charged vapor consisting of free-floating molecules, without destroying them.

John B. Fenn, a professor of chemistry at Virginia Commonwealth University in Richmond, solved this problem in 1988 by designing a syringe-like mechanism into which proteins in a water solution were

> **"It is my desire to keep walking through life as an engineer, and to continue producing results that are useful to society."**

squeezed through a needle charged by thousands of volts of electricity. As the charged droplets evaporated, the repulsion of electric charges between the molecules caused them to break apart into individual charged molecules floating in vapor, which could then be accelerated via an electric field into a detector—as per the usual mass-spectrometry technique.

Tanaka's solution to this problem was to apply a laser pulse to a protein in order to break it apart into individual, freely hovering molecules, which could then be sent through the drift space of the spectrometer. One of the great challenges Tanaka had faced was determining the matrix (a substance used to assist ionization) as well as other parameters. He used a nitrogen laser beam with a wavelength of 337 nanometers (a nanometer equals one billionth of a meter), which, as explained on the Nobel e-Museum website, "is not absorbed by the aromatic amino acids in proteins and peptides. This is important for avoiding fragmentation." Tanaka demonstrated his technique, known as soft laser desorption (SLD), at a symposium in Takarazuka in 1987; it caused a stir in the scientific community. SLD has been developed by many scientists to become known as the Matrix Assisted Laser Desorption/Ionization (MALDI) technique and is now used throughout the world to determine the mass of macromolecules. Both Tanaka and Fenn's methods have great potential to help in developing drugs to combat such diseases as malaria and ovarian, breast, and prostate cancer; the techniques are also useful for detecting harmful or cancer-causing agents in foods. Recalling his innumerable trial-and-error experiments, Tanaka noted in his autobiographical statement, "Most of the work performed by a development engineer results in failure. However, the occasional visit of success provides just the excitement an engineer needs to face work the following day."

Before receiving the Nobel Prize in Chemistry, in 2002, Tanaka's only other recognition for his work was the Research Award he received from the Mass Spectrometry Society of Japan, in May 1989; the corecipient was his associate, Yoshikazu Yoshida. "I am sure that receiving this Nobel Prize will also have an effect on my life," he noted in his autobiography. "However, just as when there is an unexpected result in an experiment which brings a pleasant surprise, I hope that I will be able to continue enjoying my life in a natural manner after receiving this prize. It is my desire to keep walking through life as an engineer, and to continue producing results that are useful to society."

Koichi Tanaka married Yuko Ikegami, who also grew up in Toyama, in May 1995. In January 2003 he was named general manager of Shimadzu Corporation's mass spectrometry research laboratory.

Further Reading

"A Noble Soul." *Shimadzu.com*. Shimadzu Corporation, n.d. Web. 9 Aug. 2012.

Chang, Kenneth. "3 Whose Work Speeded Drugs Win Nobel." *New York Times* 10 Oct. 2002: A33. Print.

"Koichi Tanaka." *Nobelprize.org*. Nobel Media, n.d. Web. 9 Aug. 2012.

Terrett, Nicholas

British pharmacologist and coinventor of Viagra

Born: February 14, 1958; Plumstead, England

Before April 1998, when the anti-impotence drug Viagra first became available in the United States, men seeking medical treatment for erectile dysfunction faced dreary prospects. The usual treatments for impotence employed an unwieldy arsenal of medical gear, including vacuum pumps, penile implants, urethral suppositories, and penis rings. But with the introduction of Viagra (the name is reportedly a portmanteau word confected by Pfizer, the drug's manufacturer, from "vigor" and "Niagara"), it became possible to treat impotence conveniently, discreetly, and with a success rate of about 70 percent. "We've always been waiting for the magic bullet," Dr. Fernando Borges, a physician who specializes in sexual dysfunctions, told Bruce Handy for *Time* (May 4, 1998). "This," he continued, "is pretty close to the magic bullet." "I've waited seven years for a real erection," one enthusiastic Viagra patient wrote, as reported by Handy. "This little pill is like a package of dynamite; you don't know if you're going to diffuse the little sucker or if it's going to explode in your face."

Viagra proved to be a cash cow for Pfizer. Within weeks of its introduction on the American market, the drug's sales had exceeded the pharmaceutical company's most optimistic forecasts. By September 1998 Viagra had become one of the bestselling drugs in the United States, with prescriptions outpacing those of both the antidepressant Prozac and the baldness remedy Rogaine. Sales of Viagra amounted to around $900 million in 2001, and combined sales of Viagra and similar anti-impotence medications marketed by other companies were expected to reach $1.8 billion by the end of 2003.

At the head of the team of research chemists who first synthesized sildenafil citrate (the scientific name for Viagra's active ingredient) and subsequently isolated its medical properties was Nicholas Terrett, a Cambridge-trained chemist employed by Pfizer, the global pharmaceutical giant. In addition to his role in the discovery of Viagra, Terret has published more than twenty technical papers, holds eleven patents, and has written a textbook, *Combinatorial Chemistry*.

Early Life and Education

Nicholas Kenneth Terrett was born on February 14, 1958, in Plumstead, a district to the south of London, England. He took an early interest in science and has recalled being fascinated with crystal gardens in junior high school. (A crystal garden is a simple chemistry experiment in which one "grows" crystals out of a saturated solution.) "I think that the crystal garden visibly demonstrates something about the variety, color, and fascination of science," Terrett explained to David Bradley for the *Alchemist*. In 1976, upon graduating from a technical high school in Bexley, Kent, he matriculated at Cambridge University's Pembroke College. Terrett earned a bachelor's degree in natural sciences from Pembroke in 1979; two years later he enrolled at Cambridge's Corpus Christi College for graduate study. In 1984 Terrett completed a PhD in organic chemistry with a doctoral thesis on the stereochemistry of the synthesis and reactions of allylsilanes.

Soon after completing his studies, Terrett took a job with Pfizer's pharmaceutical research facility in Sandwich, England. "I applied to Pfizer at the end of my PhD," he told Bradley. "I had previously been interviewed by them as an undergraduate, and knew their reputation for quality science. On that occasion they offered and I declined. We were both more lucky second time around." For his first two years at the company, Terrett led a team of chemists working to develop a compound for the treatment of heart disease.

Life's Work

It was the work on pharmaceuticals for the treatment of cardiovascular ailments that led—quite serendipitously—to the discovery of Viagra. In 1987 Terrett joined a Pfizer project to develop a lead compound aimed at reducing hypertension, or high blood pressure. Blood pressure is regulated in part by a naturally occurring body chemical known as cyclic guanosine monophosphate (cGMP), which triggers the relaxation of vascular muscles (muscles in the walls of blood vessels).

Normally, cGMP is broken down by the enzyme phosphodiesterase type 5 (PDE5). Terrett and his colleagues sought to reduce hypertension by creating a compound that would inhibit the ability of PDE5 to break down cGMP, thus leading to a relaxation of vascular muscles and, ultimately, decreased blood pressure.

Terrett and his colleagues first synthesized sildenafil citrate—along with some 1,600 related compounds—in 1989. By that time the focus

> **"Speaking from a chemist's perspective, many, many people in the pharmaceutical industry spend their entire lives failing to produce anything that gets onto the market."**

of the project had shifted from hypertension to angina, another cardiovascular ailment. Pfizer initiated clinical trials of sildenafil in 1991, and while the drug proved less effective than had been hoped in treating angina, several test subjects reported a peculiar side effect: erections, even among men previously suffering from erectile dysfunction.

That reaction piqued the curiosity of the researchers, who then set out to provide a scientific explanation for the phenomenon. Terrett and his colleagues already knew that, in a fully functioning penis, erection is caused by increased blood flow to the spongy erectile tissue that constitutes the bulk of the penis's mass. What they wanted to determine was whether erectile function (or lack thereof) was regulated by the presence of cGMP and, especially, PDE5—the enzyme known to be affected by sildenafil. Biochemical analysis of a frozen penis revealed significant quantities of PDE5, suggesting this was indeed the case. The link between PDE5 and erectile function was further tested by soaking slices of fresh (as opposed to frozen) penis in a sildenafil solution, then stimulating the nerves of the erectile tissue with electricity. As expected, the presence of sildenafil in the penile tissue suppressed the effects of PDE5; that in turn led to an accumulation of cGMP and a dilation of the blood vessels in the penis, thus paving the way for an erection.

With the drug's priapic side effect on a firm scientific basis, researchers sought to develop an oral tablet specifically for the treatment of impotence. By early 1994 Terrett and his colleagues were ready for clinical trials in the Pfizer laboratory. In the first experiment, twelve men suffering from erectile dysfunction were put on a sildenafil regimen, then fitted with a RigiScan (a laboratory device used to calibrate the plasticity and girth of a test subject's member) and exposed to various stimuli of a libidinous nature. To the researchers' delight, ten of the twelve achieved erection. "It was incredible, absolutely incredible," said Peter Ellis, Terrett's longtime colleague at Pfizer, as quoted by Rosie Mestel for *Discover* (December 31, 1998). Terrett, in an interview with *Biography* (December 1998), described the result as "probably the key moment . . . we had the first clinical evidence that we actually had an oral tablet that could be effective in treating erectile dysfunction."

Pfizer spent the next four years preparing to bring Viagra to market—a process that included further clinical studies as well as the creation of a method for the mass production of the drug. Viagra was approved by the Food and Drug Administration on March 27, 1998. "I don't think you can go through this process and not feel that you'll look back and say, 'this was the highlight of my career,'" Terrett explained to *Biography*. "Speaking from a chemist's perspective, many, many people in the pharmaceutical industry spend their entire lives failing to produce anything that gets onto the market. In fact, only the minority are fortunate enough to have a chemical compound they designed or synthesized actually reach the market. So to achieve that is significant in its own right. And very few are as successful as Viagra."

In his free time Terrett enjoys listening to classical music (especially orchestral and instrumental music of the late nineteenth and early twentieth century), playing the piano, and gardening. He is married to Sheila Frances Terrett, whom he wed in 1984. They five children, Ben, Jack, Fay, Alice, and Rebecca.

Further Reading

Houlton, Sarah. "Company Profile: Chemical Ensemble."*Chemistry World* Mar. 2011. Print.

Kenny, Ursula. "Drug Innovators: Dr. Nick Terret and Dr. Ian Osterloh, Viagra." *The Observer.* Guardian News and Media, 30 Mar. 2002. Web. 9 Aug. 2012.

Fox, Barry. "Frozen Stiff." *New Scientist* 2145 (1 Aug. 1998). Print.

Wambugu, Florence

Kenyan plant geneticist

Born: August 23, 1953; Nyeri, Kenya

Dr. Florence Wambugu has, for the past several years, been one of the most outspoken defenders of the biotechnology industry's push embracing genetically modified, or "transgenic," crops, made by altering the genetic sequence of one organism by inserting sections from the gene sequence of a different species. Wambugu herself was involved in the development of one genetically modified foodstuff, a variant of a Kenyan sweet potato that was designed to resist a recurring crop virus that could wipe out a season's crop. She has continually touted the biotechnology industry's argument that new technology could alleviate hunger by providing greater crop yields, enriched staples, and disease- and insect-resistant varieties. (Additionally, transgenic crops purportedly could prevent environmental damage by using less land for farming with the expectation of better yields.) Opponents argue that transgenic crops tamper with natural selection to such an extent that they could prove dangerous to human health and the environment, and therefore require years of testing before they can be considered acceptable. Wambugu, however, derides such caution: "You people in the developed world are certainly free to debate the merits of genetically modified foods," she told Joe Schwarcz for the Montreal *Gazette* (October 4, 2003), "but can we please eat first?"

Early Life and Education

One of nine children, Florence Wambugu was born on August 23, 1953, into a poor family in Nyeri, a small village in Kenya. She was raised on a small farm. Following Kenya's independence from Britain in 1963, Wambugu's father left the family. According to one reporter, Lynn J. Cook, in an article for *Forbes* (December 23, 2002), he "was rounded up—like so many young men—and trucked off to work on a white settler's farm." When Wambugu was thirteen, her mother sold the family cow—their most valuable possession—to raise enough money to send her to secondary school. "The reason I became a plant scientist was to help farmers like my mother," she wrote in an opinion piece for the *Washington Post* (August 26, 2001). In 1975, after completing an advanced level of high school education at the Kabare Girls High School in the nearby Kirinyaga district, she enrolled at the University of Nairobi. She graduated in 1978 with a bachelor's degree in botany and zoology. She stayed on at the university for two more years, receiving training through the botany department in tissue-culture technology—a process which duplicates plant cells in laboratories, enabling technicians to clone specific plant varieties—at the Muguga research station of the Kenya Agricultural Research Institute (KARI). She left Kenya in 1982 to attend North Dakota State University, graduating in 1984 with a master's degree in plant pathology, specializing in potato viruses.

During the following several years, Wambugu participated in numerous workshops and conferences dedicated to plant epidemiology, sweet potatoes, and gene studies, several of which were organized by the Peru-based Centro Internacional de la Papa (CIP, or International Potato Center). The CIP awarded her a grant in 1989 to study the sweet potato, which was the central crop of her mother's farm. Describing her academic pursuit, Wambugu told William Allen for the St. Louis *Post-Dispatch* (September 6, 1992) that she had become interested in the crop because "it's a high-calorie food. Americans are worried about too many calories, but not Africans. We are looking for calories." Potatoes have become a worldwide staple partially because they can be grown in dry regions and in poor soil, without fertilizers or pesticides. Between 1988 and 1991 Wambugu conducted research toward her doctorate through a joint program of KARI and the University of Bath, in which she utilized tissue-culture technology to increase the productivity of pyrethrum, a white flower in the *Chrysanthemum* genus that is fatal to insects as well as the sweet potato feathery mottle virus (SPFMV)—one of the factors that consistently decrease sweet potato yields. As a result of her studies, Wambugu was awarded the

1990 Farmers Support Award from the Pyrethrum Board of Kenya, and the following year she received a PhD in virology and biotechnology from the University of Bath.

Life's Work

In 1991 Wambugu was recruited by one of the largest biotechnology companies in the world, Monsanto, to research genetically-modified sweet potatoes at their Life Sciences Research Center, based near St. Louis, Missouri. Wambugu has explained her decision to pursue transgenic methods to develop sweet potatoes as one reached out of exasperation: "At the beginning of my career, I worked side by side with farmers in Kenya's fields trying to improve production of sweet potatoes through traditional plant breeding, . . ." she wrote for the *Washington Post*. "But after years of hard work and frustration, I finally realized I would not be able to develop a virus-resistant potato through traditional plant breeding." The US Agency for International Development (USAID) and Monsanto each contributed about $50,000 a year to Wambugu's project, which aimed to create a sweet potato genetically resistant to SPFMV. Such a potato, Wambugu and her colleagues argued, would vastly improve the sweet potato yield and could alleviate hunger on a widespread level. (SPFMV can cause the potato's growth to be stunted, and, according to Wambugu, reduces African yields of the crop by about half.) Their experiments involved inserting genetic material found in pyrethrum that made it resistant to SPFMV into the genome of the sweet potato. Monsanto allowed Wambugu to use its laboratory techniques and materials for free and stated that it would waive all intellectual property rights on any products that came from the project. In the early 1990s she framed her research as a move against cash crops which, in an earlier era, would have been more appropriate for colonial relationships. Instead of designing crops better suited for export from Africa, she was developing crops that could better feed Africa. "Real security isn't wealth," she told James E. Ellis for *BusinessWeek* (December 14, 1992). "It's food. Biotechnology can give security to poor farmers."

Wambugu returned to Kenya in 1994 to become the director of the African regional branch, or "Africenter," of the International Service for the Acquisition of Agri-biotech Applications (ISAAA), a nonprofit organization that promotes the benefits of biotechnology and works to develop transgenic crops. The ISAAA has working relationships with most of the larger biotechnology companies, including Monsanto, Novartis, and Pioneer Hi-Bred International, but it also receives funding from the New York-based Rockefeller Foundation and the Ottawa-based International Research Development Centre. Its industry ties

> **"Real security isn't wealth. It's food. Biotechnology can give security to poor farmers."**

are significant enough, however, that critics often view it as a "front" organization designed by biotechnology firms to promote their interests. As the debate over transgenic crops became more pronounced in the late 1990s, Wambugu emerged as one of the more vocal defenders of the new technology. Greenpeace and other groups skeptical of the industry push for transgenic crops argued that instead of trying to feed the hungry, multinational corporations were attempting to gain greater control of the world's food, were only interested in the profits made from monopolizing crops (which could be accomplished by patenting specific seed varieties), and, by splicing completely different gene pools together, were tampering with natural selection in an unsafe manner. In 1999 the European Union issued a moratorium on all imports of genetically modified food, which was a stumbling block for the introduction of such crops in Africa because many countries on the continent relied on the money made from exporting food to Europe. In response, Wambugu told Jeff Otieno for the Nairobi *Nation* (September 16, 1999) that "these people come from Europe and tell us biotechnology is bad but their countries have already developed capacity in the science." To Otieno, she argued that Kenya should embrace the technology to reap the short-term benefits of greater food availability, and debate on its benefits and drawbacks only afterward. Noting that farmers already use hybrid seeds—developed through the breeding of distinct but closely related varieties of plants (such hybrids usually cannot reproduce adequately), Wambugu told Otieno that "transgenic seeds are simply an added-value improvement to these hybrids." "I'm not here to fight Greenpeace or to fight for industry," she told Pauline Tam for the Ottawa *Citizen* (November 18, 1999). "I'm here to fight for Africans because we have a right to decide for ourselves." In a more conciliatory moment, she told Tam in a later article for the Ottawa *Citizen* (January 26, 2000) that "everybody's opinions need to be respected whether they are pro or against. But just to say that we'll throw away this technology would be to lose a major opportunity." Yet the accusatory nature of arguments resurfaced when she told Jeff Otieno for the Nairobi *Nation* (April 21, 2000) that "[Organizations against the use of genetically modified organisms] don't want Africa to embrace biotechnology because they know the technology has the potential to solve Kenya's famine and poverty problems." On February 14, 2000 she was announced as a member of DuPont's external Biotechnology Advisory Panel. Despite her efforts to convince skeptics of the benefits of transgenic crops, in August 2000 South Africa issued a five-year moratorium on the production, sale, and use of genetically modified food, declaring that more tests were required to accurately assess the impact of such food on the environment and on human health.

At the June 2001 Biotechnology Industry Organization (BIO) meeting, Wambugu indicated, as cited by Kathleen Hart for *Food Chemical News* (July 2, 2001) that African scientists (whom she represented) and consumers were satisfied with allowing transgenic foods to be eaten in America for several years before Africans began consuming them, adopting a more cautious attitude reflected elsewhere in the biotechnology industry. "We were blinded by our enthusiasm," Monsanto CEO Hendrik Verfaille said in November 2000, as quoted by John Vidal for the London *Guardian* (December 1, 2000). "We missed the fact that this technology raises major issues for people—of ethics, of choice, of trust, even of democracy and globalization. When we tried to explain the benefits, the science, and the safety, we did not understand that our tone—our very approach—was arrogant." At the BIO meeting Wambugu released her book, *Modifying Africa: How Biotechnology Can Benefit the Poor and Hungry, a Case Study From*

Kenya (2001), which presented many of the arguments for the adoption of transgenic crops she had stated at lectures and to journalists. Notably, at that conference, she dismissed the possibility of "gene escape"—when transgenic plants intermingle with non-transgenic ones and invade the latter's gene pool—as manageable, contrary to fears of several prominent environmentalists and agronomists. Continuing to restate her case, she wrote an editorial first published in the *Los Angeles Times* (November 11, 2001), and subsequently reprinted in numerous newspapers, lambasting anti-biotechnology protestors as an affluent minority attempting to dictate what the poor can and cannot eat, and further accusing them of using mainstream media to spread an unsubstantiated misinformation campaign.

Wambugu left her position as the regional African director of the ISAAA to found Africa Harvest Biotech Foundation International, an offshoot of the ISAAA with more of an emphasis on advocacy instead of science, in January 2002. She continued to attend conferences, deliver lectures on the benefits of transgenic crops, and give interviews and expert advice to newspapers. In 2002 Zimbabwe, Mozambique, and Zambia refused aid from the United States in the form of genetically modified food. (Zimbabwe and Mozambique eventually accepted transgenic corn after being pressured by UN relief agencies, but Zambia refused. Wambugu blamed Zambia's rejection of the aid on Europe because, based on the argument of "gene escape," Europeans could conceivably refuse exports from the country on the assumption that they could be tainted by transgenic genes.)

As Wambugu engaged on a speaking tour in parts of North America and Australia in 2003, her personal opponents became more outspoken, attacking her political and financial affiliations as well as her research: The sweet potato that she had worked on at Monsanto, which other scientists had continued to develop after she returned to Africa, had been introduced in Kenyan field tests beginning in about 2000. Despite promising results in the United States, Monsanto released reports in January 2004 that the sweet potato variety grown in Kenya was unable to satisfactorily resist SPFMV, and that the sweet potatoes in the control section of the experiment (i.e., the non-transgenic potatoes) produced a higher yield than the transgenic crops. Opponents of Wambugu's have argued that the sweet potato is a minor crop in Kenya, not considered a staple, and that 75 percent of the farmers in Uganda—where the sweet potato is a more important crop—and Tanzania already had access to non-transgenic SPFMV-resistant sweet potatoes, which could have easily been exchanged with Kenyan farmers had they sought it. Wambugu's opponents also point to studies indicating that the low sweet potato yields Wambugu cited as the average for Kenya were also drastically understated, with the real average being about the same as Wambugu's projections for the yield of her transgenic potato.

Despite the shortcomings of the sweet potato project, during Wambugu's tenure as director at ISAAA she also oversaw a project introducing tissue-culture banana trees to Kenyan farmers, which, according to the Africa Harvest website, "had a positive impact on over 500,000 small scale farmers in Kenya and the East African region" by 2003. For their research into tissue-culture banana trees, Wambugu and several colleagues won one of the two 2000 Global Development Network awards, sponsored by the World Bank, in the science and technology for development section.

In October 2002 Wambugu won the International Biographical Centre Lifetime Achievement Award. The editors of *Forbes* named Wambugu one of "15 People Who Will Reinvent Your Future" in the December 2002 issue of the magazine. In 2008, Wambugu received the Yara Prize, in recognition of her continual efforts in the Africa Green Revolution. She has written more than sixty papers for local and international journals and remains the chief executive officer of Africa Harvest Biotech Foundation International. She lives in Nairobi and has three children, Benson, James, and Marybeth.

Further Reading

Cook, Lynn J. "Millions Served." *Forbes*, 23 Dec. 2002: 302. Print.

Lruse, Loren. "Hope for a New Direction." *Successful Farming*, 1 Nov. 2005. Print.

Mabunda, Doris. "Florence Wambugu." *Contemporary Black Biography*. The Gale Group, 9 Aug. 2012.

Wambugu, Florence. "Protesters Don't Grasp Africa's Need." *Los Angeles Times* 11 Nov. 2001. M1. Print.

Wells, Spencer

American geneticist

Born: April 6, 1969; Marietta, Georgia

"The greatest history book ever written is the one hidden in our DNA," the geneticist Spencer Wells said, as posted on the website of the Genographic Project, a five-year initiative to collect 100,000 DNA samples from people around the world. Wells spearheaded the project in 2005—which is cosponsored by the National Geographic Society and IBM, as well as various private donors—in an effort to trace mankind's geographic and genetic passage to the present day. The site explains, "The fossil record fixes human origins in Africa, but little is known about the great journey that took *Homo sapiens* to the far reaches of the Earth. How did we, each of us, end up where we are? Why do we appear in such a wide array of different colors and features? Such questions are even more amazing in light of genetic evidence that we are all related—descended from a common African ancestor who lived only 60,000 years ago. Though eons have passed, the full story remains clearly written in our genes—if only we can read it." The project, while widely lauded by the scientific community, has occasionally met with negative reaction: members of some indigenous groups, for example, have characterized the DNA collection as exploitive. Others are disturbed because population genetics, as Wells's field is known, has challenged certain long-held notions of race—notions he believes are not only socially divisive and harmful, but scientifically incorrect. "People are an interwoven tapestry," Wells told Tania Tan for the Singapore *Straits Times* (July 28, 2007). "There's a little bit of everyone in everyone."

Early Life and Education

Spencer Wells was born on April 6, 1969, in Georgia and was raised in Lubbock, Texas. His mother was a biologist at Texas Tech University and his father was a tax attorney. Wells, who aspired as a young child to be a writer or a historian, credits a 1979 PBS series called

Connections for sparking his interest in science. He spent time in the lab with his mother and realized, "Science was really fun and cool. It's not just about geeky guys in white lab coats, it's about solving puzzles on a daily basis," as he recalled to Jane Gitschier for the Public Library of Science (PLoS) *Genetics Journal* (March 30, 2007). Wells attended a progressive school that grouped students according to ability rather than age; he typically studied with students older than himself and is often referred to in the media as a prodigy. At age sixteen, Wells enrolled at the University of Texas, where he majored in biology and worked on an independent study of theoretical population genetics. Elected to the honor society Phi Beta Kappa, Wells finished his undergraduate studies in just three years.

While making plans to attend graduate school at the University of Texas, Wells was invited to study with the famed evolutionary geneticist Richard Lewontin at Harvard University in Cambridge, Massachusetts. There he earned his PhD in population genetics in 1994, at the age of twenty-five. For his dissertation, Wells studied the genome of *Drosophila* (commonly called the fruit fly). He grew increasingly weary of working in a lab, however, and longed to get out in the field and study humans. "At the end of the day I wasn't terribly interested in the population structure of fruit flies," he told Gitschier, "but I had always been interested in human history." (He additionally explained that his grandfather, who died in World War II, had a reputation as a fearless military man: "I think maybe I got some of my love of danger and going to strange places from him.")

Life's Work

Wells went on to conduct postdoctoral research at the Stanford University School of Medicine in California, where he worked with Luigi Luca Cavalli-Sforza, who is credited with founding the field of human population genetics. While at Stanford, Wells became excited about the study of genomic diversity in indigenous populations as a means of tracking the migration of ancient man.

Questions of human origin and migration had long been explored almost exclusively by cultural anthropologists, who examined relationships among human populations through the disciplines of ethnography, archaeology, and linguistics. That changed greatly in the 1980s, when developments in molecular biology made possible the study of genetic variations at the DNA level. (DNA, or deoxyribonucleic acid, is the genetic material that acts as a blueprint for cells.) While 99.9 percent of DNA is common to all humans, and most of the remaining 0.1 percent is a mixture of genetic material from a given person's mother and father, there are certain genes that are passed from individual parent to offspring fully intact. Mitochondrial DNA (mtDNA), for example, is passed directly from mother to child, and the Y chromosome is passed unchanged from father to son. When DNA is replicated during cell division, a mutation occasionally occurs; if such a mutation occurs on either the mtDNA or the Y chromosome, that mutation becomes a genetic marker, and it is passed down to all subsequent generations. "If you share a marker with someone, you share an ancestor in the past," Wells explained to a journalist for the *Chicago Tribune* (April 14, 2005). "It's by looking at the pattern of these variances, and connecting people with these markers in networks, that we can trace people around the world."

Geneticists have determined that all modern humans are descended from one woman and one man—dubbed Adam and Eve—who lived in Africa between 60,000 and 200,000 years ago. The notion of our African origins is supported by tests that show "more genetic diversity in a single African village than in the whole world outside Africa," Wells told Greg Callaghan for the *Australian Magazine* (July 30, 2004). Archaeologists have determined that the first wave of humans—probably no more than a couple of thousand—migrated out of Africa to the shores of eastern Asia between 50,000 and 70,000 years ago, when major climate changes affected Africa; all of today's non-Africans share a genetic marker common to that first migrating group. Once in Asia, the population split, with some moving into the Middle East and others venturing to India. Scientists have estimated that humans first crossed the Bering Strait into the Americas some 15,000 to 20,000 years ago. "The movement was probably imperceptible," Wells explained to James Shreeve for *National Geographic* (March 1, 2006). "It was less of a journey and probably more like walking a little farther down the beach to get away from the crowd."

The Genographic Project

Wells's postdoctoral research focused on developing more detailed maps of those ancient migrations by gathering DNA directly from indigenous populations and analyzing it. The Genographic Project website explains, "In a shrinking world, mixing populations are scrambling genetic signals. The key to this puzzle is acquiring genetic samples from the world's remaining indigenous and traditional peoples whose ethnic and genetic identities are isolated." Efforts to collect DNA from indigenous populations had been stymied in the past, however, largely for geopolitical and cultural reasons. In the 1990s, for example, Cavalli-Sforza had headed the Human Genome Diversity Project, an effort to collect DNA samples from hundreds of indigenous populations; though the scientific community viewed the enterprise as universally beneficial, many indigenous groups felt exploited. Some raised concerns of intellectual property rights and requested compensation for any medical or commercial use of their DNA. Others found the act of collecting DNA samples antithetical to their cultural or religious beliefs. Still others worried that having concise information about their origins could jeopardize land rights that had been granted on the basis that their ancestors had always occupied a particular area. Because of the controversy, the US government refused to fund the project, which then came to a halt. Wells hoped that in an increasingly globalized world, his own efforts would be better received. (He took particular care to make clear that the DNA samples would not be put to any commercial use.)

In the summer of 1996, Wells traveled to the central Asian nations of Uzbekistan, Kyrgyzstan, and Kazakhstan, areas that had previously been the focus of very little genetic research. He spent five weeks collecting human blood samples from members of thirteen indigenous groups, whose ancestors had lived in the same place for centuries. Subjects received analyses of their DNA, and Wells reported that most people were pleased to participate. "People tend to get really excited about that, I find," Wells told Gitschier. "They want to know more. They say 'I'll give you the sample, but make sure you get the information back to me, and tell me what it's all about.'" Wells told Elia Ben-Ari for *BioScience* (February 1999), "[During the 1996 trip] we found evidence of what we think are Middle Eastern genes [moving] eastward along the Silk Road into China," and explained that a certain Y chromosome marker appeared in high frequency in Mongolians, but hardly at all in central Asians, suggesting that although the Mongols had conquered Asia militarily, they "probably didn't leave as strong a genetic impression as they did a cultural and historical one."

Excited by his preliminary data, in 1998 Wells decided to expand his research to include some 25,000 miles of the legendary Silk Road, an intercontinental trade route connecting the imperial court of China,

> **"The greatest history book ever written is the one hidden in our DNA."**

in the East, to the Roman Empire, in the West. Wells and his colleagues dubbed the trip Eurasia '98 and launched a website, which they updated from the road with journal entries and photographs. Starting from London, they spent six months driving across Europe through Central Asia, stopping to work with local officials and physicians to collect genetic material—about fifty samples per location. Along the way, the team encountered occasional political difficulties; one notable example took place on the border between Azerbaijan and Iran, during a time when the United States had no formal relations with the latter country. Lacking the credentials to drive across the border, the group split up, with part of the team flying to the Iranian capital, Tehran, and Wells ferrying the car across the Caspian Sea.

At the completion of the trip, Wells was appointed the director of the Population Genetics Research Group, part of the Wellcome Trust Centre for Human Genetics at Oxford University in England. "Things started to get really exciting scientifically. We had a huge number of samples," Wells told Gitschier. "Every experiment we ran was exciting and new. You're getting these results and they start to make sense. Piecing together migratory patterns." One newly discovered genetic marker that had originated in Mongolia some eight hundred years ago indicated that 8 percent of the men of central and eastern Asia—and one in 200 men in the world—is a descendant of Genghis Khan, a thirteenth-century ruler of the Mongolian Empire.

Wells next accepted a position as head researcher at a Massachusetts-based biotechnology company. While there, a producer from the Public Broadcasting Service (PBS) contacted him about making a film explaining his work. The result was *The Journey of Man: A Genetic Odyssey*, a video documentary that followed Wells and a team as they collected DNA samples throughout Africa, Australia, the Middle East, the Himalayas, and the United States. The documentary aired on PBS in the United States in 2001 and internationally on the National Geographic Channel in late 2002. Wells also wrote a companion book of the same title, published in February 2002. In a review for the *New York Times* (March 2, 2003), Carl Zimmer wrote, "Wells traces our distant history with a mix of clarity and charm that's rare among scientists. He makes the complexities of population genetics wonderfully clear with smart metaphors. And he navigates gracefully from his home waters of genetics into paleontology and climatology and back again."

"At the time I wrote the book, we had sampled maybe 10,000 people around the world, 10,000 out of 6.5 billion," Wells explained on the *Charlie Rose Show* (January 23, 2006). "That's not a great sample size." Thanks in large part to public interest in the book and film, however, the National Geographic Society offered to fund further research, and IBM and the Waitt Family Foundation signed on as well. With backers in place, Wells designed the five-year, $52 million Genographic Project. In addition to creating a vast genetic database for use by scientists, the project, which was officially launched on April 13, 2005, aims to raise awareness about the issues facing indigenous populations and to educate the public about genetics and anthropology. Wells has also invited private individuals to take part by purchasing kits priced at $199.95, which enable them to submit their DNA for testing via a cheek swab. The test results are sent to the purchasers and incorporated into the collective database (if permission is granted). "Your results will reveal your deep ancestry along a single line of direct descent (paternal or maternal) and show the migration paths they followed thousands of years ago. Your results will also place you on a particular branch of the human family tree," Wells wrote for the project website. "Your individual results may confirm your expectations of what you believe your deep ancestry to be, or you may be surprised to learn a new story about your genetic background."

A portion of the proceeds from the kits funds the Genographic Legacy Project, which awards grants for the cultural preservation of the participating indigenous groups. Like previous DNA collection efforts, however, the Genographic Project earned significant criticism from various indigenous groups. Debra Harry, the executive director of the Indigenous Council on Biocolonialism (ICBC), told Charlie Furniss for the journal *Geographical* (September 1, 2006), "All over the world we are being killed, we are being displaced. And while this is going on, the Genographic Project is spending millions of dollars on a study that hopes to show the patterns of population migrations. It's hard to see how this is a collaboration. Why don't they bring that money to us and ask us what we really need?" As had been the case with the Human Genome Diversity Project, some tribal leaders expressed concern that the project's findings on human migration would undermine their claims on their native lands and threaten their security. As before, some considered the idea of obtaining genetic material spiritually offensive and disrespectful to their ancestors, and they worried that migratory mapping might contradict beloved traditional stories about their origins. Wells has interpreted such criticisms as politically motivated and paternalistic. "We have encountered little resistance to the project in the field, where we are able to explain the project directly to prospective participants," he told Furniss. "The vast majority of those we have approached—more than 95 percent—have agreed to participate. In parts of North America and Russia, we've even had communities approaching us. Those who have refused have done so because of a fear of needles or something, not because they object to the project."

Wells—dubbed an "explorer-in-residence" for the National Geographic Society—documented many of his genetic discoveries in his second book, *Deep Ancestry: Inside the Genographic Project*, which was published in late 2007. That year Wells was the winner of the Foundation of the Future Kistler Prize, a $100,000 cash award "given to a scientist or research institution that has, with courage and wisdom, pursued the truth and made original, substantive, and innovative contributions in the study of the connections between the human genome and human society," according to the foundation's website.

A press release posted on the Genographic Project's website (July 15, 2008) announced that project researchers had discovered a previously unknown mtDNA sequence, containing a deletion—an abnormality in which part of a single chromosome has been lost—that constituted nearly 1 percent of the total mtDNA genome. That variant was unusual, because the deletion exists in a region of the genome that was previously thought to be critical for the replication process. According to the press release, "Comprising nearly 52,000 individual mtDNA genotypes from individuals from approximately 180 countries, the update represents a significant increase over the 21,000 mtDNA genotypes released by the project in 2007." Wells and other project scientists credited the unprecedented size and scope of the project's database, and the correspondingly comprehensive amount of data, for enabling them to make their discovery.

Wells has been featured in several PBS and National Geographic documentaries, including *Explorer: Quest for the Phoenicians* (2004), *The Search for Adam* (2005), and *China's Secret Mummies* (2007). He lives in Washington, DC, with his wife, the documentary filmmaker Pamela Caragol Wells. He has two daughters, Margot and Sasha, from a previous marriage. In his free time he enjoys skiing, sailing, and taking photographs, among other activities.

To date, the Genographic Project has collected more than 500,000 samples of DNA from across the world.

Further Reading

Ben-Ari, Elia T. "Molecular Biographies." *Bioscience* 49.2 (Feb. 1999): 98. Print.

Kilian, Michael, and Jeremy Manier. "Your Ancestry Might Surprise You." *Chicago Tribune* 14 Apr. 2005: C12. Print.

"Spencer Wells: 'At Root, We're Still Hunters.'" *Independent.co.uk*. The Independent, 7 June 2010. Web. 9 Aug. 2012.

Wade, Nicholas. "Still Evolving, Human Genes Tell New Story." *New York Times* 7 Mar. 2006: A1. Print.

Whitson, Peggy

American biochemist and astronaut

Born: February 9, 1960; Mount Ayr, Iowa

Peggy Whitson first imagined being an astronaut at the age of nine, when, perched eagerly before the television set, she watched Neil Armstrong and Buzz Aldrin become the first humans to walk on the moon in 1969. Thirty-three years later, on June 5, 2002, the Iowa native fulfilled her dream when she and two Russian cosmonauts launched into space on the shuttle *Endeavor*, bound for a six-month stay aboard the International Space Station (ISS). Whitson, a biochemist who worked as a research scientist for the National Aeronautics and Space Administration (NASA) for ten years before being selected as an astronaut, was only the second woman to live on the space station and the first research scientist to work in the US lab. During her first mission to the ISS, she logged 184 days, 22 hours, and 14 minutes in space as part of Expedition Five. During her voyage she wrote for the NASA Human Spaceflight website, "To be a participant in all of this is unbelievable, even to me as I float here and write this, knowing that you can see a speck of light speeding by in the early morning sky or at dusk, and knowing that I am in that bit of light. . . . It is difficult enough to comprehend the reality of this experience while I float here, and my fear is that I will not be able to hold onto the threads of this reality when I return. But I guess it will be, by far, the best dream I have ever had!"

Early Life and Education

Born on February 9, 1960, in Mount Ayr, Iowa, Whitson was raised on an eight-hundred-acre hog farm near the town of Beaconsfield, Iowa (which, according to the 2000 US Census, has a population of eleven and is the smallest town in the state). Her parents, Keith and Beth Whitson, are farmers. Her teachers remember her as a bright, dedicated student with a positive attitude. During high school she played basketball and competed in track. In 1978, the same year NASA selected its first women astronauts, she graduated second in her class from Mount Ayr Community High School.

At Iowa Wesleyan College, in Mount Pleasant, Iowa, Whitson completed a bachelor's degree in biology and chemistry in three years, graduating summa cum laude in 1981. Four years later she earned a doctorate in biochemistry from Rice University in Houston, Texas. She chose Rice for her postgraduate studies because Houston is the home of the Johnson Space Center, the primary research and training site for NASA's space program. After completing a fellowship at Rice, she became a National Research Council resident research associate at the Johnson Center. That same year she applied for the Astronaut Candidate Program; after being turned down, she applied every year thereafter for the next decade.

In April 1988 Whitson was hired as a supervisor for the Biochemistry Research Group at KRUG International, a medical-sciences contractor at the Johnson Center. The following year she began working for NASA as a biochemist, conducting experiments on the physical challenges of being in a zero-gravity environment. Three years after joining NASA, Whitson was promoted to project scientist in the Shuttle-Mir Program, through which the US government provides funding for research conducted by NASA astronauts aboard Russia's Mir Space Station. From 1995 to 1996 she served as cochair of the US-Russian Mission Science Working Group. In addition to her duties at NASA, Whitson has served on the faculties of several universities: From 1991 to 1997 she was an adjunct assistant professor in both the Department of Internal Medicine and the Department of Human Biological Chemistry and Genetics at the University of Texas Medical Branch in Galveston. She next accepted a position as an adjunct assistant professor at Rice's Mabee Laboratory for Biochemical and Genetic Engineering.

Life's Work

In April 1996 Whitson was finally accepted as an astronaut candidate. In August of that year, she began the two-year training and evaluation program NASA requires of its astronauts, after which she was assigned to technical duties at the Johnson Space Center. Meanwhile she waited to be assigned to a space flight. In 2002 Whitson was selected to be one of three astronauts to go on the fifth expedition to the International Space Station.

Since October 2000 the ISS, which is in orbit 240 miles above Earth's atmosphere, has been continually occupied by a crew of two to three astronauts, with new crews arriving every five to six months.

> **"My first impression [of Earth from Space] was of the clarity and richness of the colors that make up our planet and its atmosphere. It's like having someone turn on the lights after having lived in semidarkness for years."**

The astronauts oversee the station's ongoing construction and conduct experiments. NASA usually requires that an astronaut participate in at least one short-term flight before being assigned to a longer one, but because of Whitson's ten-year association with the space agency, that prerequisite was waived. Whitson was thrilled when she heard she was assigned to the ISS. "I don't think I would have turned down anything," she told Irene Brown for United Press International (June 14, 2002), "but my first choice was to fly on station. I want to get up there, start doing science, and help get the station constructed."

On Board the International Space Station

On June 5, 2002, Whitson and two Russian cosmonauts, Valery Korzun and Sergei Treschev, launched into space aboard the shuttle *Endeavor*. Two days later, the crew docked with the ISS, and Whitson began a six-month stay, thereby becoming the first American with a PhD, the first without a military background, the first research scientist, and the second woman to serve on the ISS. During her stay Whitson worked on two large installations for the station as part of the ongoing construction of the ISS, which is scheduled to be completed in 2013. After her promotion to science officer while aboard the ISS, Whitson oversaw twenty-five science experiments in physics, medicine, and biology. She was the ISS's first science officer, a position created in response to a report criticizing the limited amount of scientific research conducted on the station. One of the main experiments she handled involved growing soybeans as part of a study on the effects of zero gravity on plant growth and genetics. "They looked so good, Sergei thought we should eat them as a salad," she told Julie Bain for *Ladies' Home Journal* (April 2003). "I managed to stop him and save the science. We'll need this knowledge because if we go to Mars one day, we'll need a garden onboard!"

The highlight of Whitson's trip was a four-and-a-half-hour walk in space, during which she installed micrometeoroid panels on the exterior of the ISS to protect it from space debris. On the SpaceRef website (August 27, 2002), Whitson likened the experience to flying, but noted that it was "more like the one you have in your dreams when you fly from place to place without the aid of any craft, only the view from space was much better than anything I had ever dreamed of." Describing the view of Earth, she told Bain, "My first impression was of the clarity and richness of the colors that make up our planet and its atmosphere. It's like having someone turn on the lights after having lived in semidarkness for years."

Whitson, Korzun, and Treschev returned to Earth on December 7, 2002, after being repeatedly delayed by shuttle fuel-line cracks, an oxygen leak, robot-arm problems, and bad weather. Upon her return she told a crowd of NASA staffers, as quoted in the *Houston Chronicle* (December 10, 2002), "I want you to know the space station is so much more spectacular than any video, than any photo, than any words we could ever say. I want you to be really proud of your accomplishments."

On February 1, 2003, two months after Whitson's return, the shuttle *Columbia* exploded after reentering the atmosphere following a mission in space, killing all seven astronauts onboard. All shuttle launches were halted pending the completion of an investigation into the causes of the disaster. Whitson has said that in spite of the tragedy, she is eager to return to space. "Space exploration is part of us as human beings, and I think we have to continue exploration," she told Chris

Clayton in an interview for the *Omaha World Herald* (April 11, 2003). "I'm ready to go back into space as soon as they let me."

Whitson has received numerous awards, the most recent of which are the American Astronautical Society Randolph Lovelace II Award (1995) and the Group Achievement Award for the Shuttle-Mir Program (1996). She holds two patents. In June 2003 Whitson led a crew of three in a sixteen-day training program in the underwater laboratory *Aquarius*, off the coast of Key West, Florida. The program is designed to accustom astronauts to the close quarters, isolation, and risks of space flight. In 2006, Whitson received the NASA Outstanding Leadership Medal and, in 2009, she chaired the Astronaut Selection Board. She served as backup ISS commander for Expedition 14, and returned to the ISS as station commander for Expedition 16 from October 2007 to April 2008.

Personal Life

Whitson and her husband, Clarence F. Sam, a research scientist who also works at the Johnson Space Center, live in Houston; they have no children. In her spare time Whitson enjoys windsurfing, biking, basketball, water skiing, and, especially, gardening. "If she couldn't have been an astronaut, she would have been a farmer," Whitson's aunt, Ann Walters, told Ken Fuson for the *Des Moines Register* (June 5, 2002). "She's still the same old Peggy we've always known. She's still just a farm girl from back here."

Further Reading

Carreau, Mark. "Astronaut Chronicles Her 5-Month Voyage." *Houston Chronicle* 10 Nov. 2002: A1. Print.

"Commander Peggy Whitson Breaks Record for Time in Space for a U.S. Astronaut." *ScienceDaily*. ScienceDaily, 21 Apr. 2008. Web. 9 Aug. 2012.

Munson, Kyle. "America's Uncertain Mission." *Des Moines Register*. Gannett, 28 Jan. 2011. Web. 9 Aug. 2012.

Wüthrich, Kurt

Swiss chemist

Born: October 4, 1938; Aarberg, Switzerland

On October 9, 2002, the Royal Swedish Academy of Sciences awarded Kurt Wüthrich the Nobel Prize in Chemistry for his contributions to the field of nuclear magnetic resonance (NMR). Wüthrich shared the prize with John B. Fenn and Koichi Tanaka, who made advancements in the related field of mass spectrometry (MS), an analytical method used to identify a molecular substance in a sample by accurately determining its molecular mass. While NMR and MS have been in existence for some time, enabling scientists to study small- and medium-size molecules, the work of Wüthrich and his colaureates made it possible for scientists to use these methods on large biological macromolecules for the first time.

Protein structures derived using NMR have proven useful to the pharmaceutical industry in designing drugs to treat such illnesses as cancer and HIV. In its press release, the Nobel committee praised Wüthrich and his colaureates for pioneering discoveries that have "revolutionized the development of new pharmaceuticals." Wüthrich has determined more than fifty novel NMR structures of proteins and nucleic acids, including the immunosuppression system cyclophilin A-cyclosporin A and the homeodomain operator DNA transcriptional regulation system. He has determined numerous protein pheromone structures found in Mediterranean Sea creatures and pheromone-binding proteins found in such organisms as silkworms. Wüthrich's work has also been important in the study of the murine, human, and bovine prion proteins. Although prion proteins are a regular component of the brain tissue of all mammals, scientists believe that misfolded prions are the cause of bovine spongiform encephalopathy (BSE), also known as mad cow disease, as well as Creutzfeldt-Jakob disease, which afflicts humans.

Early Life and Education

Kurt Wuthrich was born on October 4, 1938, in Aarberg, Switzerland. He studied chemistry, physics, and mathematics at the University of Bern, Switzerland, from 1957 to 1962. He then studied physical education at the University of Basel, Switzerland, where he received a PhD in 1964. He was a postdoctoral fellow at Basel until 1965, and then at the University of California, Berkeley until 1967. He worked for two years as a member of the technical staff at Bell Telephone Laboratories in Murray Hill, New Jersey, before joining the Swiss Federal Institute of Technology (Eidgenössische Technische Hochschule or ETH) in Zurich, Switzerland, in 1969. He became an assistant professor there in 1972, an associate professor in 1976, and a professor in 1980. From 1995 to 2000 he chaired the biology department; he is currently a professor of biophysics. Since 2001 he has divided his time between the ETH Zurich and The Scripps Research Institute (TSRI), where he is the Cecil H. and Ida M. Green Visiting Professor of Structural Biology. (Wüthrich's award marked the second year in a row that Scripps scientists have shared in the Nobel Prize in Chemistry. K. Barry Sharpless is a 2001 laureate.) Wüthrich is also the deputy director and a principal investigator in the Zurich-based National Center of Competence in Research Structural Biology (NCCR Structural Biology).

The Study of Protein Structure

Wüthrich and his team began working toward the NMR method for protein-structure determination in the mid-1970s, at a time when scientists were still largely at a loss about how to use this technology to analyze large molecules in solution. Biological macromolecules, such as deoxyribonucleic acid (DNA), proteins, and complex sugars, are the key components of the cells of living organisms. While DNA provides the blueprint for the composition of the cell (through protein synthesis), it is the proteins that carry out important cellular activities such as growth and cell maintenance. Each protein has a particular biological function; hemoglobin, for example, is the protein that carries oxygen to all the cells of the body. Understanding proteins and their role in cellular life is, accordingly, a high priority for biologists.

The function of a protein is largely determined by its shape, but even the largest proteins are far too small to be studied at sufficient resolution under a microscope. Scientists have therefore devised other ways to create pictures of protein structures. The first protein structures studied at atomic resolution were myoglobin and hemoglobin by Max Perutz and John Kendrew in the late 1950s. Perutz and Kendrew, who were awarded the Nobel Prize in 1962, used a method known as X-ray crystallography, which, in the decades following their discovery, became the primary method of studying protein structures. X-ray crystallography involves carefully growing protein crystals (which can be extremely time-consuming) and then bombarding them with X rays. Scientists can then study the diffraction pattern of the X rays to deduce the structure of the protein.

X-ray crystallography is limited because it cannot be used on molecules in solution, which is closer to the way proteins are actually found in their natural states in living systems. To complement X-ray crystallography, scientists developed NMR spectroscopy. The technique makes use of the fact that some atomic nuclei, because of a physical property known as their "spin," absorb radio-frequency waves and, in response, emit radio-waves. In an NMR experiment, a molecular sample in a glass tube is inserted into an NMR magnet and subjected to a range of frequencies. When an external magnetic field of the appropriate frequency is introduced, the magnetic fields of the spin particles are disturbed out of their alignment, causing them to emit radio-waves. The radio-wave responses of the atoms are measured and recorded, revealing an NMR spectrum that is unique for each type of molecule. The spectrum can then be used to reconstruct the shape of the molecule.

NMR phenomena were first detected in 1946 by the Swiss physicist Felix Bloch and the US physicist Edward Purcell. In the early 1950s it was discovered that NMR signals carried information about the chemical environment of the nucleus studied, prompting the development of NMR as an analytical tool in chemistry. Its use was limited, however, as it required incredibly concentrated solutions. In 1966 the Swiss chemist Richard Ernst found that the sensitivity of NMR could be increased using Fourier transform spectroscopy, by exposing the sample to short, intense radio-frequency pulses instead of the previously used weak "continuous wave" (CW) irradiation. Ernst thus opened the way for a wide range of novel applications of NMR in chemistry, biology, and medicine.

Although NMR proved very useful for small and mid-size molecules, it did not work well for large molecules like proteins. Protein

> **"I think it's one of the basic human rights to have access to modern science."**

molecules contain hundreds to thousands of atoms, making it difficult for scientists to interpret the countless radio-waves that are emitted. "It gets to be a very complex problem," Dr. Adriaan Bax, the section chief for biophysical nuclear magnetic resonance spectroscopy at the National Institutes of Health, told Kenneth Chang for the *New York Times* (October 10, 2002). "It's essentially a big jigsaw puzzle," Bax explained. "It's a 10,000-piece puzzle with all the same color pieces."

Life's Work

Wüthrich was able to overcome the problem "by careful bookkeeping," as he told Chang. He devised the sequential-assignment method, which has become the cornerstone of all NMR structural investigations of macromolecules. Sequential assignment involves a systematic method of pairing each NMR signal with a particular hydrogen nucleus in the macromolecule. The signals are then evaluated to determine the distances between pairs of hydrogen atoms. By applying a mathematical formula, these distances can then be used to calculate the three-dimensional structure of the molecule (in the same way that, if one has the measurements of a house, one can draw a three-dimensional image of the house). By noting the structural variations of a given protein, scientists are also able to understand its dynamics, or how it interacts with other molecules in solution.

Having conducted studies in such areas as spin diffusion in proteins and the sequential-assignment strategy for proteins in the 1970s, in 1982 Wüthrich and his team devised the framework for NMR structure determination of proteins and published a series of influential papers outlining their research. The first complete determination of a protein structure using Wüthrich's method came in 1985. Since then the use of NMR in protein analysis has blossomed. Although the majority of known protein structures have been obtained using the traditional X-ray crystallography techniques, by May 2002 about 20 percent of the known protein structures had been determined using NMR in solution.

Wüthrich has authored more than 600 research papers and several books, including *NMR of Proteins and Nucleic Acids* (1986), the definitive book on the subject. He also travels around the world to discuss his research with students living in countries where scientific research lags behind more developed nations. "He believes he has a social obligation" to aid young scientists in the developing world, remarked a friend and colleague at the University of California, San Diego, as quoted by Bruce Lieberman for the San Diego *Union-Tribune* (October 10, 2002). Wüthrich explained to Lieberman, "I think it's one of the basic human rights to have access to modern science . . . and that these students have a chance to learn about what is going on."

Wüthrich holds honorary doctorates from twelve universities and has received numerous honors and awards for his work, including the Louisa Gross Horwitz Prize (1991), Switzerland's Marcel Benoist Prize and the Louis Jeantet Prize for Medicine (1993), the Kyoto Prize in Advanced Technology (1998), and the President's Gold Medal (2011) from the government of India. He is a member of the European Molecular Biology Organization and of several national and local academies in Europe, the Americas, and Asia. He is also a foreign associate of the National Academy of Sciences in the United States and a foreign honorary member of the American Academy of Arts and Sciences. In 2010, Wüthrich was elected as a foreign member to the Royal Society in the United Kingdom. He is a former secretary general and vice president of the International Union of Pure and Applied Biophysics (IUPAB) and a member of the General Committee of the International Council of Scientific Unions (ICSU). He has held numerous endowed lectureships and has had visiting faculty appointments at major universities in the United States, Japan, and Europe.

Wüthrich and his wife, Marianne Briner, have two children, Bernhard Andrew and Karin Lynn.

Further Reading

Chang, Kenneth. "3 Whose Work Speeded Drugs Win Nobel." *New York Times* 10 Oct. 2002: A33. Print.

Maugh, Thomas H. II. "3 Win Nobels in Chemistry for Molecular Studies." *Los Angeles Times* 10 Oct. 2002: I31. Print.

Wüthrich, Kurt. "Autobiography." *Nobelprize.org*. Nobel Media, 2002. Web. 10 Aug. 2012.

Zewail, Ahmed H.

Egyptian American chemist

Born: February 26, 1946; Damanhur, Egypt

The Egyptian American chemist Ahmed H. Zewail was awarded the 1999 Nobel Prize in Chemistry for his groundbreaking work in the field of femtochemistry. Using ultrashort pulses of laser light, he was able to visually record for the first time the actual bonding or cleaving of molecules during a chemical reaction. His pioneering investigations of fundamental chemical reactions and the methods he used to perform them have led to explosive developments in research that have a far-reaching affect on all branches of science.

Early Life and Education

The chemist Ahmed H. Zewail was born on February 26, 1946, in Damanhur, Egypt. He began his professional career in 1966 as an undergraduate trainee at the Shell Corporation in his native country. A year later he received his bachelor of science degree from Alexandria University in Egypt. He continued his studies at Alexandria, earning a master's degree in 1969, while working there as a chemistry instructor and researcher. In the early 1970s Zewail traveled to the United States to study at the University of Pennsylvania in Philadelphia, where he earned his PhD in chemistry in 1974. From 1974 to 1976 he worked at the University of California at Berkeley on an IBM postdoctoral research fellowship. In 1976 he joined the California Institute of Technology (Caltech) as an assistant professor of chemical physics. Here, over the next two decades, he would conduct his pioneering research in the field of femtochemistry—the study of chemical acts that occur in a femtosecond, or one-quadrillionth of a second. (A femtosecond is equal to 0.000000000000001 of one second, which is to 1 second what 1 second is to 32 million years.)Femtochemistry was once a rarely explored field. Before Zewail began his work, scientists were unable to watch the breaking or making of molecules, since no camera was fast enough to capture these actions. All cameras were limited by the time an electric current could travel through the wire and semiconductor materials that comprise switches. No electronic circuit could be built that could turn the flash circuit off and on faster than one-hundredth of one-billionth of a second—far slower than a femtosecond. Knowing these limitations, scientists had to settle for before-and-after pictures of the creation or destruction of molecules, which left the actual point

> **"Look at the world. Everything around us is chemical reactions. Everything—inside you and me, the atmosphere, everything we breathe, we touch."**

of bonding or cleaving unrecorded. In order to capture that moment in which molecules bond or cleave, Zewail had to rethink the way in which that instant could be visually recorded.

Life's Work

Zewail found his answer in mode-locking, a technique developed by physicists at Bell Laboratories in the early 1980s that worked at the femtosecond level. According to Gary Taubes, who wrote about Zewail's work for *Discover* (February 1994): "To understand mode-locking . . . start with a light bulb. The light emitted is composed of electromagnetic waves, all with random wavelengths and out of phase, meaning that they're oscillating up and down out of step with one another. The result is the familiar beam of white light. A laser beam, in contrast, generates all the light waves at virtually a single wavelength, or color, and all in phrase, which is to say they're 'coherent,' all oscillating in more or less perfect step." However, an ordinary laser produces a continuous electromagnetic wave, which is unsuitable for what Zewail wanted to accomplish. He needed something more akin to a strobe effect: coherent pulses of light flashing in a femtosecond.

Zewail's solution was to generate light at more than one wavelength—10,000 different wavelengths, in fact. Each wavelength would be evenly spaced but out of phase 99.996 percent of the time. In phase the other .0004 percent of the time, the wavelengths would line up exactly, with all the peaks and troughs level. A single pulse of laser light would thus be created, enabling the movement of the molecule to be photographed as if with an incredibly fast shutter. By the late 1980s his research group was able to create a kind of motion picture of chemical reactions by taking a series of frames of the reaction.

One of the earliest discoveries Zewail's work uncovered, according to a report in *BBC News* (October 12, 1999), "was the realization that as chemical reactions proceed intermediate products are formed that are quite distinct from the reactants and final products." Since their first experiments, Zewail and his associates have studied more than fifty molecular reactions and are branching out into different fields. In the subbasement of Noyes Hall on the Caltech campus, the group works in an area dubbed "Femtoland," in which five laboratories are devoted to different types of chemical reactions, examining everything from the simple breaking and bonding of molecules to how the dynamics of a tangle of forces on a complex molecule affect it. Zewail's work has greatly altered academic research in chemistry labs across the United States, moving it away from Bunsen burners and beakers and toward advanced laser technology on a microscopic level. Such labs are busy experimenting on the dynamics of proteins, as well as the mechanics of human eyesight. The significance of this work is not lost on the man who has been dubbed the "father of femtochemistry" by the American Chemical Society. "Look at the world," Zewail told Gary Taubes. "Everything around us is chemical reactions. Everything—inside you and me, the atmosphere, everything we breathe, we touch. Everything is a chemical reaction. So we have to develop a unified theory of how chemistry takes place. We can't understand this unless we really have a coherent understanding of how atoms and molecules like or dislike each other. That's our ultimate goal."

As the pioneering figure of femtochemistry, Ahmed Zewail has been greatly honored for his work, both in the United States and abroad. In addition to winning the 1999 Nobel Prize for Chemistry, Zewail has won the Buck-Whitney Medal from the American Chemical Society (1985), the King Faisal International Prize in Science (1989), the Nobel Laureate Signature Award (1992), the 1993 Earl K. Plyler Prize from the American Physics Society, the 1993 Wolf Prize in

Chemistry, and the 1993 Medal of the Royal Netherlands Academy of Arts and Sciences, among others. In 1990 he was named Caltech's first Linus Pauling Professor, a position established in memory of the two-time Nobel laureate. Zewail has been twice honored by the country of his birth: first in 1995 by former Egyptian president Mubarak, who presented him with the Order of Merit, First Class, and again three years later when the Egyptian government issued two postage stamps with Zewail's portrait. In April 2009, US president Barack Obama appointed Zewail to the President's Council of Advisors on Science and Technology and, six months later, Zewail was named an envoy in the newly established US Science Envoy Program, which was created to foster collaboration between scientists worldwide. In 2011, the American Chemical Society selected Zewail to receive the Priestley Gold Medal, in recognition of "his development of revolutionary methods for the study of ultrafast processes in chemistry biology, and materials science."

Ahmed H. Zewail, who holds citizenships in the United States and Egypt, lives in San Marino, California, with his wife Dema Zewail, a physician in public health at the University of California, Los Angeles. They have four children.

Further Reading

Barth, Amy. "Scientist of the Arab Spring." *Discover Magazine*. Kalmbach, 3 Jan. 2012. Web. 9 Aug. 2012.

Gedik, Nuh. "Science in the Islamic World: An Interview With Nobel Laureate Ahmed Zewail." *The Fountain* 67 (January 2009). Print.

Zewail, Ahmed. "Autobiography." *Nobelprize.org.* Nobel Media, Jan. 2006. Web. 9 Aug. 2012.

Appendixes

Historical Biographies

Amedeo Avogadro

Italian physicist

A pioneer in atomic theory, Avogadro was the first scientist to distinguish between atoms and molecules. Avogadro's law, a hypothesis that relates the volume of a gas to the number of particles present, greatly advanced the understanding of chemical reactions and resolved many chemical problems.

Areas of Achievement: Chemistry, physics

Born: August 9, 1776; Turin, Kingdom of Sardinia (now in Italy)

Died: July 9, 1856; Turin, Kingdom of Sardinia (now in Italy)

Early Life

Amedeo Avogadro (AH-vah-GAH-droh) was born in Turin in the Kingdom of Sardinia about 60 miles southwest of Milan in what is now Italy. He was one of four sons born to Count Filippo Avogadro and Anna Maria Vercellone. Count Avogadro was a distinguished lawyer and civil servant who came from a prominent family in the region that had produced many generations of Italian military and civil administrative leaders. The name Avogadro possibly is derived from the Italian word *avvocato* (barrister).

As a young child, Avogadro likely received his first education at home from the local priests; he later attended secondary schools in Turin. Between 1792 and 1796 he studied law at the University of Turin with the intention of following his father in a legal career. For some years after graduating from law school, he held several government positions. Around 1800 he began to show an interest in natural philosophy, undertook private study of physics and mathematics, and attended physics lectures at the university. His interest in science seems likely to have been stimulated by the recent research on electricity by fellow Italian Alessandro Volta, who came from neighboring Lombardy.

After 1806, Avogadro abandoned his interest in a legal career to concentrate on science and, with one of his brothers, began working on electricity experiments. He was soon appointed as a demonstrator at the Academy of Turin. In 1809 he became professor of natural philosophy at the Royal College of Vercelli. Within one decade, Avogadro was elected as a full member of the Turin Academy of Sciences and one year later was appointed to the first Italian chair of mathematical physics at Turin. His salary was six hundred lire per year.

Life's Work

Avogadro was a prolific writer and published articles in many areas of the physical sciences throughout his life. His name appears in most modern chemistry and physics textbooks, although he has often been misrepresented as being a chemist because his work had a profound influence on the development of chemical theories. His name is usually associated with two important aspects of chemistry: Avogadro's law, which describes the relationship between the volume and number of particles of a gas, and the Avogadro number, which represents the number of particles in one mole of a substance. The concept of the mole as a unit for the measurement of atomic particles was unknown in Avogadro's time, and the term was not introduced until the twentieth century.

In many ways it is remarkable that Avogadro had a successful scientific career. He received no formal training in science and made a dramatic career change when he was thirty years old. The former was not particularly unusual in the eighteenth and nineteenth centuries, as some of the greatest scientists of the period were self-educated, including Humphry Davy and Michael Faraday. Others (for example, Nicolas Lémery and Jakob Berzelius) had received their early training in a related field such as pharmacy or medicine before concentrating on a career in the physical sciences.

It was surprising that Avogadro made the transition from law to science so effortlessly, and it was an obvious testament to his good mind and dedicated spirit of discovery. However, Avogadro was not a good experimentalist and had a poor reputation as such among his colleagues. He preferred to interpret the experimental results of others using a mathematical approach. Much of his work was translated and published, but it generally appeared in obscure journals. In addition, Turin was geographically isolated from the world centers of scientific research, which were generally considered to be in Germany and France. Finally Avogadro was, by nature, modest and reserved, and he never actively sought fame. He never traveled to other countries and rarely corresponded or met with other scientists outside his region. It was not until after his death that the world really comprehended and recognized his contributions to science.

During the early nineteenth century, chemists began serious attempts to understand the nature of matter and chemical reactions. John Dalton measured the mass ratios of elements in compounds and found these ratios to always be simple whole numbers. For the first time, he demonstrated that the elements must exist as discrete units, or atoms. The nature of one particular form of matter, gases, had always been difficult for early scientists to understand. In 1808, Joseph-Louis Gay-Lussac published studies on the combining volumes of gases. He showed that gases always combined in simple whole number ratios. For example, 200 cubic centimeters of hydrogen always combined with 100 cubic centimeters of oxygen to form 200 cubic centimeters of water vapor (a 2:1:2 ratio).

Although such observations suggested that equal volumes of gases contained equal numbers of atoms, Dalton rejected this hypothesis, believing that Gay-Lussac's experiments were inaccurate. Dalton and others argued that one volume of oxygen gas contained a specific number of oxygen atoms and therefore must produce the same volume of water vapor with an equivalent number of water atoms. It should be remembered that at the time, it was still generally believed that water had a chemical formula of HO and was composed of HO atoms.

Like Dalton, most chemists of the day believed that common gaseous elements such as hydrogen, oxygen, nitrogen, and chlorine existed as individual atoms. Avogadro's explanation appeared in his 1811 article *Essai d'une manière de déterminer les masses relatives des molécules élémentaires des corps et les proportions selon lesquelles elles entrent dans ces combinaisons* (essay on a manner of determining the relative masses of the elementary molecules of bodies and the proportions in which they enter into combinations), in which he attempted to explain the inconsistencies with existing theories by assuming that equal volumes of all gases contained equal numbers of molecules rather than atoms, provided conditions of temperature and pressure were kept constant. Avogadro's hypothesis (also known as the molecular hypothesis) would later become known as Avogadro's law. During a chemical reaction, therefore, Avogadro proposed that molecules could split into half-molecules (atoms) and combine with other half-molecules to form the observed product compounds.

By contrast, Dalton viewed the combination of two gases such as hydrogen and oxygen as involving individual atoms. It is now known, thanks to Avogadro's insight, that this reaction involves molecules that are composed of two atoms each (diatomic molecules), which is consistent with Gay-Lussac's experimental results on combining volumes in which two volumes of hydrogen react with one volume of oxygen to generate two volumes of water vapor. According to Avogadro, each molecule of water must contain one molecule of hydrogen (H_2) and one half-molecule of oxygen (one O atom). It followed from this that the correct chemical formula for water was H_2O and not HO as Dalton and others believed. It should be noted that Avogadro never used the modern system of chemical formulas that are shown in the previous equations. If he had, his theory may have been more understandable and therefore readily accepted sooner.

Avogadro's hypothesis also explained discrepancies in the measured densities of gases and resulted in more accurate determinations of atomic weights. Water vapor was known to have a lower density than oxygen, but this fact was difficult to explain if the latter existed as single atoms. The occurrence of diatomic oxygen molecules easily explained why oxygen had a greater density than water vapor. In Dalton's early table of atomic weights, which were a measurement of the relative weights of atoms, hydrogen was assigned a value of 1 and oxygen a value of 7.5. When viewed as diatomic molecules, the value of hydrogen became 2 and oxygen 15. In other words, Dalton's atomic weights had to be doubled for diatomic molecules. This eventually led to a more accurate table of atomic weights.

Avogadro's molecular hypothesis was largely ignored during his lifetime. His theory did have the support of a fellow Italian chemist, Stanislao Cannizzaro, who was one of the few who seemed to grasp the significance of Avogadro's idea, but only after Avogadro's death. Most scientists, however, failed to distinguish between atoms and molecules, and Avogadro, isolated in Turin and largely unknown in Europe, never witnessed the universal acceptance of his theory. Cannizzaro showed that Avogadro's theory could be used for determining molecular size and accurate chemical formulas. His enthusiasm for the molecular hypothesis had a profound influence on the German chemist Lothar Meyer. In his 1864 textbook, Meyer employed Avogadro's hypothesis to develop his ideas on theoretical chemistry. This book had considerable influence on other chemists, who applied Avogadro's ideas to many other aspects of physical chemistry.

Avogadro held his position as the chair of mathematical physics at Turin from 1820 until 1822, when it was abolished because of regional political turmoil. The position was reestablished in 1832, and Avogadro was reappointed in 1834. He held this post until his retirement in 1850 at the age of seventy-four. He spent the last six years of his life continuing with his scientific studies and died in Turin on July 9, 1856.

Significance

It was not until around 1870 that the term "Avogadro's law" first appeared in print; by the 1880s it had received universal recognition. The realization that common elemental gases existed as diatomic units had an enormous influence on obliterating chemical inconsistencies and linking the chemical and physical properties of substances. Accurate density determinations and atomic and molecular weight measurements for gases also became possible, which aided the rapidly developing area of organic chemistry in the nineteenth century.

Once Avogadro's law was understood, a new era in the development of chemical theories and molecular composition became possible. The Dutch chemist Jacobus van't Hoff showed that Avogadro's law could be applied to solutions as well as gases, for which he was awarded the first Nobel Prize in Chemistry in 1901. For chemists, a significant consequence of Avogadro's law was the realization that one mole of all substances (that is, the atomic or molecular weight of a substance expressed in grams) contains the same number of particles. This quantity, equal to 6.02252×10^{23}, is now known as the Avogadro number in honor of a great scientist who was not recognized in his own lifetime.

Nicholas C. Thomas

Joseph Black

Scottish chemist

A pioneer of quantitative experimental chemistry, Black discovered carbon dioxide, the first gas to be isolated and have its properties systematically identified. He also proposed the theories of latent and specific heats and, as a gifted lecturer, raised the profile of chemistry to a philosophical and public science.

Areas of achievement: Chemistry, science and technology

Born: April 16, 1728; Bordeaux, France

Died: December 6, 1799; Edinburgh, Scotland

Early Life

Joseph Black's father, John Black, came from a family of Scottish-Irish merchants. During the first half of the eighteenth century, John Black prospered as a factor in the wine trade. With success came prominence in Bordeaux society, and among the family's closest friends was Montesquieu, president of the Sovereign Court of Bordeaux and author of *The Spirit of the Laws* (1748). Joseph's mother, Mary, also was from a family of merchants. She had descended from the Gordons of Aberdeenshire, Scotland, and her lineage, like her husband's, connected the world of eighteenth century trade to the world of the eighteenth century Enlightenment. Among Black's cousins on his mother's side was the Scottish social philosopher Adam Ferguson, author of *An Essay on the History of Civil Society* (1767). Joseph Black would later be best man at Ferguson's wedding, and Ferguson an early biographer of Black.

Joseph Black was the ninth of fifteen children. Protestant outsiders in Catholic France, the large family compensated with close ties of mutual support. Both John Black and Mary Gordon generously provided for their children's advancement, and an extensive family correspondence reveals that the children amply returned their affection. In 1740, Joseph Black was enrolled in a private school in Belfast, where he learned sufficient Latin and Greek to enter the University of Glasgow in the fall of 1744.

By the spring of 1748, Black had completed the undergraduate arts curriculum. The following fall he began the study of medicine. Black's choice of medicine satisfied his father's concern that his son get a professional education. It also satisfied Black's own developing interests in natural philosophy. The choice proved fortunate. The previous year William Cullen had become professor of medicine at the University of Glasgow. At Glasgow, Cullen introduced a chemistry that was less subordinate to the pharmaceutical needs of medicine and more an autonomous science posing its own distinct questions—a chemistry, in short, that was more "philosophical." The young Black quickly attracted the attention of the new professor of medicine. By 1749, Black had become Cullen's laboratory assistant.

Life's Work

In 1752, Joseph Black left Glasgow to do his medical thesis at the more prestigious University of Edinburgh. Published in June, 1754, the thesis was a milestone in Scottish philosophical chemistry. In June of 1755, Black presented his results to the Philosophical Society of Edinburgh. In the now classic paper, "Experiments upon Magnesia Alba, Quicklime, and Some Other Alcaline Substances" (pb. 1756), Black demonstrated that when heated, alkaline substances such as magnesium carbonate emitted a gas that Black called "fixed air." This demonstration meant that "air" was not itself an element but instead made up of chemically distinct gases. It also meant that, contrary to previous understanding, a gas could combine with a solid. Furthermore, Black established that the gas he had discovered, fixed air, or carbon dioxide, was a by-product of fermentation and respiration, and that among its properties were mild acidity, a greater density than common air, and its not supporting life or combustion.

Black's paper secured his scientific reputation and future career. When in 1755 Cullen became professor of chemistry at the University of Edinburgh, Black succeeded him at Glasgow as a professor of medicine and a lecturer in chemistry. Black now followed Cullen's suggestions and turned to the study of heat. Black's experiments on alkalis had been distinguished by their elegant design and quantitative rigor. So, too, were his observations on heat. Furthermore, historians of science sometimes credit him with the development of the ice calorimeter. Black's observations soon led him to discover both latent heats and specific heats. The amount of heat absorbed or released by substances—say, water in freezing or water in vaporizing—without a change in temperature is a latent heat, and the different amounts of heat required to raise equal masses of different substances an equal interval in temperature are called specific heats.

Black never published his work on heat, but he did publicize it. As the rich collection of student notes makes clear, Black began his course on chemistry with a detailed presentation of his research on the effects of heat. Indeed, from the early 1760s, Black largely abandoned the laboratory for the lecture hall. In 1766, with Cullen's appointment as a professor of medicine, Black took up the chair of chemistry at Edinburgh. In his new position, Black increasingly focused on applied chemistry.

Already at Glasgow, Black had forged a close professional and, eventually, personal and business relationship with James Watt, who had been appointed instrument maker to the university. In 1769, Black loaned Watt the money needed to obtain a patent on his steam engine. As Watt himself affirmed, the methodological and theoretical background for his invention was laid by Black's meticulous program of experimentation and his investigation of latent and specific heats. Now, in the 1770s and 1780s, Scottish agricultural improvers such as Henry Home, Lord Kames, sought Black's chemical imprimatur for their proposals even as entrepreneurs sought Black's advice on the metallurgy of coal and iron, the bleaching of textiles, and the manufacture of glass. Black also maintained a medical practice and in 1776 was the attending physician at the death of philosopher David Hume.

Black's dedication to bringing together university and industry as well as philosophy and improvement was a trait he shared with the profusion of clubs in eighteenth century Scotland. He was a member not only of the Philosophical Society (later Royal Society) of Edinburgh but also of less formal civic groups, such as the Select Society or the Poker Club. Dearest to Black was the Oyster Club, weekly dinners with his closest friends William Cullen, the geologist James Hutton, and Adam Smith, the author of *The Wealth of Nations* (1776). Black's friendship with Smith went back to their days as students and then new faculty members at the University of Glasgow. When Smith died in 1790, Black and Hutton served as his executors. In the mid-1790s, Black's own health, which had never been robust, began to fail. On December 6, 1799, just three days after he had managed to complete yet another course of lectures, Black died peacefully at his home in Edinburgh.

Significance

Joseph Black was an important precursor of the chemical revolution of the late eighteenth century. Britain's preeminent professor of chemistry, Black was uniquely influential. Across his career, he introduced Scottish philosophical chemistry to as many as five thousand students. From the last decade of the eighteenth century until well into the nineteenth century, these students would edit chemical journals in Germany and hold chairs in chemistry at universities such as Cambridge, Oxford, Columbia, Princeton, and Yale.

During the 1790s, Black was among the first to bring Antoine-Laurent Lavoisier's reform of chemical nomenclature to an English-speaking audience. However, in 1789, the very year in which Lavoisier's *Traité élémentaire de chimie* (*Elements of Chemistry, in a New Systematic Order, Containing All the Modern Discoveries*, 1790) was published, Lavoisier wrote Black acknowledging "the important revolutions which your discoveries have caused in the Sciences." Black's discovery of carbon dioxide opened the way for Joseph Priestley's discovery of oxygen and Lavoisier's own demonstration of oxygen's role in calcination and combustion. To be sure, Black's measurement of latent and specific heats still worked within a framework of a chemical theory that imagined heat as a caloric fluid that was lost when a body was cooled and gained when a body was heated. His measurements, however, also turned on a distinction between heat and temperature that laid the groundwork for nineteenth century thermodynamics and a conception of heat as kinetic energy.

From his earliest years in Bordeaux, Black stood at the crossroads between the great economic transformation that was the Industrial Revolution and the great cultural transformation that was the European Enlightenment. His life work balanced a calling to make the study of chemistry "philosophical" with a civic commitment to the "improvement" of Scotland. The balance that Black achieved was a model for chemistry's continuing career in Britain as a public science—a career that culminated in Sir Humphry Davy's lectures to the Royal Institution in the early nineteenth century. The balance that Black achieved also marked a critical moment in European cultural history, a moment before specialization would estrange the "two cultures" of the sciences and the humanities, a moment when chemistry remained a liberal vocation.

Charles R. Sullivan

Robert Boyle

Irish scientist

Boyle discovered Boyle's law, which describes the relationship between air pressure and volume. He promoted the experimental method in scientific study, especially in the field of chemistry.

Areas of achievement: Chemistry, physics, science and technology

Born: January 25, 1627; Lismore, County Waterford, Ireland

Died: December 31, 1691; London, England

Early Life

Robert Boyle was the seventh son and the fourteenth child born to his parents. Boyle's father, Richard Boyle, was the earl of Cork, reported to be the wealthiest man in the British Isles at the time of Robert's birth. The earl's wife, Katherine Fenton, was the only daughter of Sir Geoffrey Fenton, the secretary of state for Ireland. Sadly, Robert Boyle hardly knew his mother, for she died of tuberculosis at the age of forty-four in February of 1630, when he was only four years old.

The wealth, prestige, and character traits Boyle inherited from both of his parents helped immensely in his career as a scientist. Most contemporaries of the Boyle family remark on the physical and temperamental resemblances between young Boyle and his father. Each was a tall man of wiry build with a long, thin, pale face and large eyes. Also, each was an intellectual and scholarly man who brought unusual amounts of energy and determination to the tasks he undertook. Richard Boyle had migrated to Ireland from England in 1588; he rose quickly in jobs of state in Ireland and made his fortune from land purchases. He made many friends and allies in public life—a phenomenon that would be repeated in his son's life. Each Boyle was also scrupulous about recording the details of his business, and each was an industrious worker.

Boyle was reared in the Irish countryside of Munster County, where his father had large landholdings. A peasant woman served as his nurse, and her cottage was his home during infancy. It was hoped that the simple foods, fresh air, and exercise that Robert had as a youngster would lead to good health in his adult years. He remained a frail if active person, however, all of his life. He was especially uncomfortable in cold weather and wore a series of cloaks to keep himself warm. At age twenty, he developed what was diagnosed as a kidney stone, a painful ailment. He spent much of his adult life seeking remedies for the stone and other illnesses. He often made his own medicines and tested their effectiveness on himself.

Boyle was strong enough at age eight to be sent to Eton in England to study along with his older brother, Francis Boyle. Robert's tutor at Eton, John Harrison, was highly influential in the boy's early intellectual development. Harrison first instilled in young Robert an immense passion for reading and learning that would remain with Boyle throughout his lifetime; this passion led him to read almost all day long. Since Boyle's eyes were weak, Harrison had to force him to leave his studies and play outdoors for a part of each day.

The Boyles had been without the services of Harrison at Eton for about a year when, in October of 1638, Richard Boyle sent his two sons to study in Europe. Their new tutor was a Frenchman they called Monsieur Marcombes; they lived with him for the next six years. The majority of this time the trio spent in Geneva, Switzerland, with some extended visits to Italy, especially to Rome and Florence. During these six years, Robert studied a variety of subjects, including French, mathematics, and theology. Boyle became so good at conversing in French that he could pass as a native Frenchman.

At age thirteen, in 1640, during a sudden and violent thunderstorm in Geneva, Robert believed that he was truly converted to a fervent Christianity. He was a devout believer for the rest of his life, and his studies in the physical sciences were always conducted so as to demonstrate the existence of God in the universe. The regularity of physical and chemical laws convinced Boyle that an intelligent God had created the world.

In conjunction with his religious beliefs, Boyle studied several ancient languages in order to read the Bible in its original form. These language studies especially occupied his teenage years. At fifteen, Boyle read and admired the works of Galileo, which would influence his later scientific studies. Boyle also admired Francis Bacon, the English essayist and philosopher who advocated a form of empiricism. In this system, the facts or data are observed in order to reach a theory rather than a theory being used to judge the data.

Life's Work

The civil wars in England and Ireland delayed Boyle's return home from his studies in Europe. At last, in the summer of 1644, he landed in England, where he fortunately found his older sister, Katherine, Lady Ranelagh, with whom he resided for several months. When his finances were set in order, he moved in March of 1646 to Stalbridge, England, where he spent the next six years reading and writing while maintaining his estate.

Boyle at this time began meeting regularly with a group of scientists of varied interests in London; this group became known as the Invisible College, since they had no permanent meeting place. From the influence of these men, Boyle became interested in experimental philosophy. He especially enjoyed studies in chemistry, since they were closely allied with his interest in medicine and remedies for illness.

Since Boyle's home at Stalbridge was not convenient to any major college, he consented to move to Oxford in the summer of 1654 when asked to do so by Dr. John Wilkins, the warden of Wadham College. In Oxford, Boyle set up an elaborate laboratory for scientific research on High Street. He was ably assisted by various aides, craftsmen, and secretaries, the most noted and skilled of these being Robert Hooke. This life of experimenting, recording, and discussion that Boyle established on High Street would continue for the next fourteen years. An immense amount of scientific and naturalistic data was recorded by Boyle and his staff during these years. Most of this information would be published for use by other scientists and scholars.

Historical Biographies 239

The first important book that Boyle published was *New Experiments Physio-Mechanicall, Touching the Spring of the Air and Its Effects* (1660). The experiments described here (in detail, as Boyle always did) were in large part related to physics rather than chemistry. Boyle had learned of an air pump invented in 1654 in Germany by Otto von Guericke. This pump created a vacuum in which scientific experiments could be performed and new physical concepts tested; it needed improvement, however, to work efficiently and consistently.

Boyle's ingenious assistant, Hooke, designed and built an improved air pump by 1659. With this new laboratory instrument, Boyle conducted numerous experiments on air pressure. From the detailed notes he kept on his work, Boyle was able to conclude in *New Experiments Physio-Mechanicall, Touching the Spring of the Air and Its Effects* that air volume always varies in inverse proportion to pressure. He further explained this important phenomenon in the second edition (1662) of the book. This physical law, known as Boyle's law, is his most widely recognized permanent contribution to the field of physics.

Boyle's other writings would have effects on the chemical sciences in addition to physics. In his *Experiments and Considerations Touching Colours* (1664), he accurately described the physical phenomena that caused the colors black and white. This book had an impact on Sir Isaac Newton's later works in optics. The next year, Boyle published *New Experiments and Observations Touching Cold* (1665), in which he discussed how thermometers could be improved. (They were in his day a fairly new and crude piece of laboratory equipment invented by Galileo.) Boyle also studied the effects of freezing on various chemical mixtures; he was able to disprove the widely held idea that water and other liquids contract when frozen. These areas of study were also later reinvestigated by Newton.

Of all of his numerous scientific publications, Boyle is most famous for his book *The Sceptical Chymist* (1661, rev. 1679). In this work, Boyle refuted two influential schools of scientific thought popular in his era: the Aristotelian and the Paracelsian. The two schools of thought were similar in that both held that all matter was composed only of a certain few basic substances (and nothing else). The Aristotelians believed that air, earth, fire, and water were the four elements that made up all animate and inanimate things in nature, while the Paracelsians held that there were three basic principles that combined to form all matter: sulfur, salt, and mercury.

Boyle demonstrated, by experimentation (something the other two systems did not employ), that these theories were incorrect. Instead, he argued that the universe was composed of numerous small particles, or corpuscles, of various shapes and sizes. Their alignments and combinations, and even their motion, determined what element would be formed. Boyle's conception of matter is known as corpuscular philosophy; it was greatly influenced by the mechanistic theory of Francis Bacon.

In the summer of 1668, Boyle returned to live in London at Pall Mall with his sister, Katherine, Lady Ranelagh. The two kept a house together, for Boyle never married, and his sister was separated from her husband. Boyle and Katherine were very popular and famous persons in their era. They entertained frequently, and Boyle even had to post a sign on their door to state when visitors were not allowed. Only in this way could he conduct his studies; he also maintained an extensive correspondence with scientists throughout England and Europe. Boyle continued his scientific work even after suffering a stroke in June of 1670, which left him partially paralyzed. He died on December 31, 1691, exactly one week after his sister had died. Both were buried in London.

Significance

Boyle made two important contributions to science, particularly physics and chemistry. The first is a tangible contribution: He was a founding member of the Royal Society of England in 1662. This group of eminent scientists, including William Brouncker, Robert Murray, Paul Neile, John Wilkins, and Sir Christopher Wren, met regularly to present papers on scientific thought and to discuss the results of experiments. The society also kept meticulous records of their proceedings (perhaps a direct result of Boyle's participation) and published the first science journal in England. The Royal Society promoted excellence and dedication in the sciences and attracted such men as Sir Isaac Newton to careers in science. Boyle, solidly a member of the British aristocracy and noted for his hard work on behalf of experimentation, did much to raise the science of chemistry to a high level of respectability (something it previously lacked in English society and culture). King Charles II officially chartered the Royal Society in 1662 and himself did some amateur experiments.

Boyle's second contribution is less concrete but nevertheless vital. Boyle's experimental approach to chemistry helped to bring it into the realm of modern scholarship. Previously, a mystical or mysterious element was associated with chemistry. Alchemy, or the alleged changing of one substance into another (most often a base metal into a precious one), was almost the only chemical investigation done until Boyle's day. Fraudulent claims, farfetched speculation, and outright trickery made alchemy an unrespectable method of study. By replacing quasi-scientific work with the experimental method, Boyle did a great service for future generations of chemical researchers.

Ironically, Boyle made no specific discoveries that remain as part of modern chemistry, although he is frequently called the father of modern chemistry. In his experiments with air, he came close to discovering oxygen, but Joseph Priestly would actually do that many years later. Similarly, Boyle's work on the nature of colors was less influential than that of Newton, who worked with a prism in his experiments. Boyle himself acknowledged his lack of acumen in experiments requiring complex mathematical calculations. He thought that his greatest contribution to future scientists would come from the mass of naturalistic data he had accumulated in his lifetime. That data, if one includes his published scientific treatises in it, did prove to be Boyle's most important legacy. He provided for others an essential foundation on which to build their own scientific accomplishments.

Patricia E. Sweeney

Henry Cavendish

French-born English scientist

Cavendish, a reclusive character, made significant advances in the chemistry of gases and contributed to the study of electrical phenomena.

Areas of Achievement: Chemistry, physics
Born: October 10, 1731; Nice, France
Died: February 24, 1810; London, England

Early Life

Little is known in detail about the personal life of Henry Cavendish. He was born into a leading aristocratic British family. His father, Lord Charles Cavendish, was the third son of the duke of Devonshire, while his mother, Lady Anne Grey, was the daughter of the duke of Kent. His mother's death when he was two years old left him totally in his father's care. He entered Dr. Newcombe's Academy in Hackney in 1742 and matriculated at St. Peter's College, Cambridge, in 1749, leaving in 1753 without taking his degree. Most biographers have speculated that he refused the religious tests required for a degree, but no evidence exists concerning his religious convictions at any time in his life.

After leaving Cambridge, he resided with his father in London. Lord Charles Cavendish was a longtime member of the Royal Society and an avid investigator of meteorological and electrical questions, having received the society's Copley Medal for perfecting a registering thermometer. His father sponsored his election to the Royal Society in 1760; living in an all-male environment permeated with scientific conversation may very well have shaped both Cavendish's lifelong fascination with science and his strange social behavior.

Throughout his life, Cavendish was reclusive, shunning human society. He was restricted by his father to an extremely small allowance, forcing him to become very close with money, often attending Royal Society dinners with only the 5 shillings necessary to gain admission. Following his father's death in 1783, he came into a large family inheritance. While very wealthy, he remained parsimonious in his personal expenditures and oddly indifferent to other uses of money, giving generously to charity but always checking the list of donors and giving precisely the amount of the largest donation. Abhorring any meeting with women—he even left notes for housemaids to avoid personal contact—Cavendish remained an eccentric and elusive person. Only in the Royal Society, where he attended regularly and participated fully, did he have any public life. He was responsible for a detailed description and analysis of the society's meteorological instruments in 1776 and served on a committee investigating lightning protection for the Purfleet powder magazine, using his own electrical research to recommend pointed rather than blunt lightning rods. Even at the Royal Society, however, he would flee if approached by a stranger, and he was often seen outside the meeting room, waiting for the moment when he could slip in unnoticed. To the end of his life he was so totally devoted to science that everything else was secondary.

Only one portrait of Cavendish exists, a watercolor sketch by Sir John Barrow done surreptitiously at a dinner meeting of the Royal Society. It shows a somewhat tall, lanky middle-aged man of a rather sharp and thrusting appearance, dressed in a greatcoat and three-cornered hat stylish in the 1750s. Apparently the dress was his usual, as he was noted for never changing the style of his clothing and purchased only one set of clothing at a time, following a precise schedule as the old garments were worn out.

Life's Work

In his lifetime, Henry Cavendish was known primarily for his research in chemistry and electricity, work reflected in a remarkably small number of papers in the *Philosophical Transactions of the Royal Society*. His first public work was in 1776, a series of papers on the chemistry of "factitious airs," or gases. Chemistry at the time was dominated by the idea that air was a single element, one of the four Greek elements. British chemists, from Robert Boyle, who first elucidated the gas laws in 1662, through Stephen Hales and Joseph Black in the eighteenth century, had made "pneumatic chemistry," manipulating and measuring "air" in its various states of purity, practically a national specialty.

Chemical research was also carried on around the organizing concept of the phlogiston theory put forward by Georg Ernst Stahl in 1723. Phlogiston was the element of fire, or its principle, which caused inflammability when present in a body. It was believed to be central to most chemical reactions. Combustion was explained as a body releasing its phlogiston. In this dual context of pneumatic chemistry and phlogiston theory Cavendish presented his study of "factitious airs," or gases contained in bodies. Most important, he isolated and identified "inflammable air," now called hydrogen. Recognizing the explosive nature of "inflammable air," Cavendish went on to identify it as phlogiston itself. He cannot be said to have discovered hydrogen, as others had separated it before him, and he did not specifically claim its discovery.

In 1783, Cavendish, having heard of experiments by Joseph Priestley that generated a "dew" upon exploding "inflammable air" and "common air," presented a study of an improved eudiometer, or a device for testing the "goodness" of air. He demonstrated that "common air" was composed of constant proportions of different constituents, rather than being a single elemental substance. This research was followed the next year with two more papers on air, the research for which was principally concerned to find the cause of the absorption of gases in the eudiometer. When he mixed "inflammable air" with "common air," he created an explosion and he noticed a dew on the containing vessel. Where Priestley had mentioned the fact in passing, Cavendish focused on it, noting that "by this experiment it appears that this dew is plain water, and consequently that almost all the inflammable air, and about one-fifth of the common air, are turned into pure water." He had discovered that water was a compound of "airs," not one of the four elements.

Cavendish's priority of the discovery of the composition of water was soon disputed. He had done his experiments in the summer of

1781 but postponed publication because the resulting water was contaminated with nitric acid. Solving this problem eventually led him to his last chemical discovery, the isolation of nitric acid in 1785. He had told Priestley of his water results, and a friend had passed the same information to Antoine-Laurent Lavoisier in Paris, who repeated and extended the experiments, while James Watt also made the discovery independently. When all three published their investigations in 1784, Watt, Lavoisier, and Cavendish all laid claim to priority. A brief controversy ensued but was quickly extinguished, with each of the three rather politely deferring to the others. The controversy was rekindled in the mid-nineteenth century and continues in some scholarly circles. Cavendish should probably be conceded the right to claim the discovery of the compound nature of water, although his explanation was given in the context of the phlogiston theory. Lavoisier followed Cavendish chronologically but explained the composition of water in the radically different terms of his new antiphlogiston chemistry, as the union of oxygen and hydrogen.

Lavoisier's antiphlogiston explanation was indicative of a revolution in chemistry that he was leading on the Continent. In 1772, when he had weighed the product of calcination (oxidation in the new terminology), there was a weight gain in the calx. He offered the explanation that something was taken up in the process, rather than phlogiston being given off. This "something" he identified as oxygen and thereby created a new chemistry. Cavendish recognized that Lavoisier's oxygen-based chemistry was essentially equivalent to a phlogiston-based chemistry, but he rejected the new ideas to the end of his life. "It seems," he wrote, "the phaenomena of nature might be explained very well on this principle, without the help of phlogiston; . . . but as the commonly received principle of phlogiston explains all phaenomena, at least as well as Mr. Lavoisier's, I have adhered to that." In 1787, Lavoisier introduced his new chemistry in his *Nomenclature chimique*, and fully elaborated it in 1789 in *Traité élémentaire de chimie*. Again Cavendish rejected the ideas, stressing that arbitrary names and ideas, as he saw them, could only lead to great mischief. By 1788, Cavendish had published the last of his chemical researches.

The second major area of Cavendish's published research was in electrical phenomena, a near-craze sweeping through scientific and popular circles throughout Europe. In 1771, Cavendish published his first paper on electricity, putting forward a single-fluid theory of electricity, in opposition to the then popular two-fluid theory. He presented electricity as a single "fluid" that could be measured according to its compressibility, using the analogy of Boyle's gas law. He provided a quantitative measure of tension as well as quantity, surmising that this "fluid" followed an inverse square law for repulsion, rather than the first power law displayed by gases. This work was expanded in 1776, with a paper describing his effort to construct a model of the electrical torpedo fish to analyze whether its electrical effects were similar to electrostatic phenomena, then the center of scientific attention. He demonstrated that they were the same.

In 1783, he also published a series of papers on heat, focusing on the question of the freezing point of mercury. He was able to expand upon the idea of latent heat presented earlier in the century by Joseph Black. These papers were based on a series of experiments carried out at the request of Cavendish by officers of the Hudson's Bay Company in 1781 and after in northern Canada.

Finally, in 1798 he presented his final paper to the Royal Society, outlining his effort to measure the density of the Earth. Using an experimental torsional balance apparatus devised by John Michel, he was able to provide an extremely accurate estimate of Earth's density, while also providing the first experimental demonstration of Sir Isaac Newton's gravitational laws.

Significance

Henry Cavendish was widely recognized by his contemporaries as a precise experimenter and a thorough researcher, despite his small output of published works. His international reputation was certified in 1803 by election as a foreign member of the Institut de France (French Institute). This reputation was further honored in the founding of the Cavendish Society in 1846 and his family's endowment of the Cavendish Laboratories and the Cavendish Professorship at Cambridge in

1871, where much of the fundamental research in atomic theory was carried out in the early twentieth century.

Unpublished manuscripts, discovered after his death, have significantly altered the perception of his scientific pursuits as limited solely to precise and narrow experimentation. The publication of these papers has deepened the appreciation of Cavendish's interests and abilities. Russell McCormmach has shown that Cavendish's work was rooted in a coherent and consistent vision of the scientific approach and theories of Newton. Newton had postulated a world made up of interacting particles, attracting and repelling one another according to strict mathematical laws. Cavendish pursued an integrated research program attempting to apply the Newtonian insight to the physical world.

Whether because he was simply shy about publishing or because he refused to publish until he was fully satisfied with the results, the world saw only a seemingly unrelated series of papers. Yet each, whether on "airs," heat, electricity, or the density of Earth, was firmly rooted in his Newtonian vision of the world. Cavendish was the most original and creative physical theorist of his age in a nation of empirical experimenters, yet he kept his ideas hidden behind a facade of the misanthropic rejection of humanity.

William E. Eagan

John Dalton

English chemist and physicist

Dalton's work with gases led him to develop modern atomic theory, and for his contributions he is considered one of the founders of the modern physical sciences.

Areas of Achievement: Physics, chemistry

Born: September 6, 1766; Eaglesfield, Cumberland, England

Died: July 27, 1844; Manchester, England

Early Life

John Dalton, born September 6, 1766, in Eaglesfield, Cumberland, was the second son of Joseph Dalton, a poor weaver, and Deborah Greenup, a woman of vigor and intelligence. His parents came from old Quaker stock, and they had six children, three of whom, Jonathan, John, and Mary, lived to maturity. John was deeply influenced by his mother's tenacity and frugality. The Society of Friends, which formed a strong social fabric in west Cumberland, was another powerful influence. The Cumberland Quakers emphasized both general education and particular training in natural philosophy, and this provided a favorableF environment for John's development as a scientist.

Dalton made rapid progress under John Fletcher in the village school, and he quickly attracted the attention of Elihu Robinson, a prominent Quaker naturalist who became Dalton's patron and lifelong friend. Because of the poverty of his family, Dalton was forced to work for a time as a farm laborer, but in 1781 he was liberated from this way of life by an invitation to replace his elder brother as assistant at Kendal, a boarding school some forty miles from Eaglesfield. The school, newly built by Quakers, had a well-stocked library that contained Sir Isaac Newton's *Philosophiae Naturalis Principia Mathematica*

(1687; *Mathematical Principals of Natural Philosophy*, 1729) as well as works of both British and Continental natural philosophers. As at Eaglesfield, so at Kendal, Dalton was fortunate in forming an important friendship. In this case, his patron and friend was the blind natural philosopher John Gough. Under Gough's tutelage, Dalton made rapid progress in mathematics, meteorology, and botany. In imitation of his master, Dalton started in 1787 keeping a daily meteorological record, a practice he continued until the day of his death.

At Kendal, Dalton began giving a series of public lectures in physics and astronomy. As a teacher, he was clear and orderly, though rather colorless. In physical appearance, he was a tall, gaunt, and awkward man with a prominent chin and nose, and he dressed in the Quaker fashion: knee breeches, gray stockings, and buckled shoes. Though modestly successful at Kendal, he became restless and sought a different profession. He made inquiries about studying medicine at Edinburgh, but he met with discouraging replies. Eventually, Dalton accepted a position as professor of mathematics and natural philosophy in Manchester. He was pleased with this appointment because, in addition to mathematics and natural philosophy, he was allowed to teach chemistry.

In 1800, encouraged by his success in Manchester, Dalton decided to resign his position and open his own "mathematical academy," where he would offer courses in mathematics, experimental philosophy, and chemistry. This endeavor prospered, and within two years Dalton had enough students to provide him with a modest income. Private teaching on this scale would occupy and support him for the rest of his days.

Life's Work

Dalton was deeply influenced by the British tradition of popular Newtonianism, a way of visualizing the world through the internal makeup of matter and the operation of short-range forces. As shown so well by Newton, these forces could be described mathematically. Besides his interest in theoretical Newtonian physics, Dalton was involved with more practical concerns—constructing barometers, thermometers, rain gauges, and hygrometers. Dalton produced essays on trade winds, proposed a theory of the aurora borealis, and advanced a theory of rain.

Dalton's meteorological investigations led him to question how the gases of the air were held together: Were they chemically united or were they physically mixed together in the air just as sand and stones were in the earth? He concluded that gases, composed of particles, were physically mixed together, and this led him to deduce that in a mixture of gases at the same temperature, every gas acts independently (Dalton's law of partial pressures). It is ironic that in trying to provide a proof for his physical ideas, Dalton discovered the chemical atomic theory. What started as an interest in meteorology ended up as a new approach to chemistry.

To support his theory of mixed gases, Dalton experimented on the proportions of the different gases in the atmosphere. It was this investigation which accidentally raised the whole question of the solubility of gases in water. In 1802, he read a paper to the Manchester Literary and Philosophical Society in which he proposed that carbon dioxide gas (which he called "carbonic acid" gas) was held in water, not by chemical attractive forces but by the pressure of the gas on the surface, forcing it into the pores of the water.

This explanation provoked William Henry, Dalton's close friend, to begin a series of experiments to discover the order of attractions of gases for water. He eventually found that, at a certain temperature, the volume of a gas taken up by a given volume of water is directly proportional to the pressure of the gas (Henry's law). Dalton was quick to see the relevance of Henry's results to his own ideas. He saw the absorption of gases by water as a purely mechanical effect. He realized that the greatest difficulty with the mechanical theory of gas-water solubility came from different gases obeying different laws. Why does water not admit into its bulk every kind of gas in the same way? Dalton answered this question by saying that gases whose particles were lightest were least absorbable and those with greater weight were more absorbable. Therefore it is clear that Dalton's important decision to investigate the relative weights of atoms arose from his attempt to find experimental support for his theory of mixed gases. The paper that he wrote on this subject in 1803 closed with the very first list of what would come to be called atomic weights.

Dalton's method of calculating the relative weights of atoms was quite simple. Following the postulates of his theory of mixed gases, he assumed that when two elements come together in a chemical reaction, they do so in the simplest possible way (the rule of greatest simplicity). For example, Dalton knew that water was a compound of oxygen and hydrogen. He reasoned that if water was the only compound of these two elements that could be obtained (and it was at the time), then water must be a binary combination of one hydrogen atom and one oxygen atom (HO). Dalton also gave rules for deciding on formulas when there were two or more compounds of two elements. Armed with this mechanical view of combining atoms, it was easy for Dalton to argue from the experimental knowledge that, in forming nine ounces of water, eight ounces of oxygen combined with one ounce of hydrogen to the statement that the relative weights of their atoms were as eight to one.

When Dalton published his table of atomic weights in a Manchester journal, his theory, for the most part, provoked little reaction. It is true that Humphry Davy at the Royal Institution rejected Dalton's ideas as speculations that were more ingenious than important. Despite this lack of enthusiasm, Dalton continued to develop his theory, and in 1804 he worked out the formulas for different hydrocarbons. By 1807, he was spreading the news of his system of chemical philosophy (the first part of which was published in 1808, the second part in 1810). With this publication, the chemical atomic theory was launched. Unfortunately, the theory was the climax of Dalton's scientific creativity, and although he did much work in several scientific fields for the next twenty-five years, the main thrust of his work was in providing atomic weights for known chemical compounds, a problem that would plague chemists for the next fifty years.

Dalton's scientific studies were nourished by the Manchester Literary and Philosophical Society. This group gave him encouragement, an audience, and recognition. In contrast, the Royal Society was dilatory in making him a member, which they did in 1822 at the urging of some of Dalton's friends. In 1831, Dalton helped to found the British

Association for the Advancement of Science, and he chaired several of its committees. His activity in these scientific societies was halted in 1837 by two severe paralytic attacks, which left him an invalid for the remaining years of his life.

Dalton lived according to regular and rigid habits. He never married, but he was deeply attached to several relatives and associates, in particular, Jonathan, his brother, and Peter Clare, his closest friend. In his later years, Dalton was admired by his countrymen, and after he died on July 27, 1844, he was given a civic funeral with full honors. His body lay in state in the Manchester Town Hall for four days while more than forty thousand people filed past his coffin. This response to his death was an indication of his scientific achievements and a manifestation of the love for him felt by the inhabitants of the city where he spent his happiest years.

Significance

John Dalton sometimes used the word "atom" in the sense of the smallest particle with a particular nature. If the atom were divided any further, it would lose its distinguishing chemical characteristics. Dalton also began, however, to think of atoms in the more radical sense of indivisible particles. With this usage, he provided a new and enormously fruitful model of matter for chemistry, because he was able to develop a way of determining atomic weights.

In his *New System of Chemical Philosophy* (1808, 1810), Dalton emphasized that chemical analysis is only the separating of ultimate particles, and synthesis is only the uniting of these particles. God created these elementary particles, and they cannot be changed. All the atoms of a particular element are alike, but the atoms of different elements differ. Though the atoms cannot be changed, they can be combined. Water, Dalton wrote, is composed of molecules formed by the union of a single particle of oxygen to a single ultimate particle of hydrogen.

Many chemists were unwilling to adopt Dalton's chemical atoms. Others used the atomic theory in a pragmatic way, for it helped to make sense of their observations in the laboratory, but they could not grasp its philosophical basis. Some scientists continued to object to atoms well into the twentieth century. They were a minority, however, and a characteristic theme of nineteenth century chemistry was the triumphant march of Dalton's ideas.

Robert J. Paradowski

Josiah Willard Gibbs

American physical scientist

Gibbs established the theoretical basis for modern physical chemistry by quantifying the second law of thermodynamics and developing heterogeneous thermodynamics. This and other work earned for him recognition as the greatest American scientist of the nineteenth century.

Areas of Achievement: Chemistry, physics
Born: February 11, 1839; New Haven, Connecticut
Died: April 28, 1903; New Haven, Connecticut

Early Life

Josiah Willard Gibbs, later known usually as J. Willard Gibbs—to distinguish him from his father, who bore the same name—was the fourth child and son among five children of J. W. Gibbs, a professor of sacred literature at Yale College Theological Seminary, and Mary Anna Van Cleve Gibbs. Born in New Haven, Connecticut, Gibbs would live all of his life in that city, leaving only to take one trip abroad, and dying in the same house in which he grew up.

Well educated in private schools, Gibbs was graduated from Yale College in 1858, receiving prizes in Latin and mathematics. In 1863, he took his doctoral degree in engineering at Yale—one of the first such degrees in the United States; his dissertation was entitled *On the Form of the Teeth of Wheels in Spur Gearing*. For the next three years he tutored at Yale in Latin and natural philosophy, working on several practical inventions and obtaining a patent for one of them, an improved railway brake. The foregoing points to a not inconsiderable practical element in the chiefly theoretical scientist that Gibbs would become.

In 1866, Gibbs embarked on a journey to Europe, where he attended lectures by the best-known mathematicians and physicists of that era, spending a year each at the Universities of Paris, Berlin, and Heidelberg. What he learned at these schools would form the basis for his later theoretical work. His parents having died, and two of his sisters as well, he traveled with his remaining sisters, Anna and Julia, the latter returning home early to marry Addison Van Name, later librarian of Connecticut Academy, in 1867.

Upon his return to New Haven in 1869, Gibbs began work at once on his great theoretical undertaking, which he would not complete until 1878. He lived in the Van Name household, as did his sister Anna; neither he nor she was ever married. In 1871, he was appointed professor of mathematical physics at Yale—the first such chair in the United States—but without salary. He was obliged to live for nine years on his not very considerable inherited income. When The Johns Hopkins University, aware of the significance of his work, offered him a position at a good salary, Yale decided to offer Gibbs two-thirds of what Johns Hopkins would pay; it was enough for Gibbs.

In 1873, he published his first paper, "Graphical Methods in the Thermodynamics of Fluids," which clarified the concept of entropy, introduced in 1850 by Rudolf Clausius. The genius of this insight was immediately recognized by James Clerk Maxwell in England, to whom Gibbs had sent a copy of the paper. The work was published, however, in a relatively obscure journal, *The Transactions of the Connecticut Academy of Arts and Sciences*, where almost all of Gibbs's subsequent writings would appear. In addition, his style was so terse, austere, and condensed as to be unreadable to all but a few readers who were already well acquainted with his underlying assumptions. Consequently, for most of his life Gibbs would remain largely unknown, especially in the United States, except among a small circle of his scientific colleagues.

This undeserved obscurity never seemed to trouble Gibbs. By all accounts, he was a genuinely unassuming and unpretentious man, tolerant, kind, approachable, and seemingly unconscious of his intellectual eminence. He was by no means gregarious and probably more than a little aloof; he had few really close friends, though he kept up a large correspondence. He attended church regularly. Physically, he was of slight build and owed a certain frailty in health to a severe case of scarlet fever in childhood. However, he was strong enough to ride and was known as a good horseman. Photographs of him reveal a handsome but somewhat stern man with a well-trimmed, short beard. The photographic image leaves an apt impression of what he was in life: a gentleman, a professor, and a scientist of unimpeachable integrity.

Life's Work

Gibbs published yet another paper in 1873, "A Method of Geometrical Representation of the Thermodynamic Properties of Substances by Means of Surfaces." In 1876, the first 140 pages of his major work appeared (again in *Transactions*), the final 180 pages finally being published by that journal in 1878; both parts bore the title "On the Equilibrium of Heterogeneous Substances." This work, of the utmost importance to science, never appeared in book form in English in Gibbs's lifetime. Its significance was appreciated by Maxwell, who incorporated some of its findings into his own books, but he died in 1879.

Continental Europeans perceived the general importance of Gibbs's discoveries, but had real difficulty reading Gibbs's text. (Gibbs himself rejected all suggestions that he rewrite his treatise as a readable book.) Hermann Helmholtz and Max Planck both duplicated some of Gibbs's work, simply because they did not know of it. A German translation of it, by the great scientist Wilhelm Ostwald, appeared only in 1892. French translations of various sections of the treatise were published in 1899 and 1903. Meanwhile, scientists came to perceive—in the words of physics professor Paul Epstein—that a

young investigator, having discovered an entirely new branch of science, gave in a single contribution an exhaustive treatment of it which foreshadowed the development of theoretical chemistry for a quarter of a century.

Gibbs's was thus an achievement almost unparalleled in the history of science. Ostwald predicted that the result of Gibbs's work would determine the form and content of chemistry for a century to come—

and he was right. A French scientist compared Gibbs, in his importance to chemistry, with Antoine Lavoisier. It should be mentioned that the editors of *The Transactions of the Connecticut Academy of Arts and Sciences* published Gibbs's work on faith alone as they were not able to understand it completely; they obtained the money for publishing the long treatise through private subscription.

Of special importance in Gibbs's work is its sophisticated mathematics. It is therefore not possible to summarize his discoveries in a brief article. There are two features, however, that must be noted. First, Gibbs succeeded in precisely formulating the second law of thermodynamics, which states that the spontaneous flow of heat from hot to cold bodies is reversible only with the expenditure of mechanical or other nonthermal energy. Consequently, entropy (S), equal to heat (Q) divided by temperature change (T), must continually be increasing. Prior to Gibbs, thermodynamics simply did not exist as a science.

Second, Gibbs derived from his more complex heterogeneous thermodynamics the "phase rule," which shows the relationship between the degrees of freedom (F) of a thermodynamic system and the number of components (C) and the number of phases (P), so that $F = C + 2 - P$. He showed how these relationships could be expressed graphically, in three dimensions. Often phase-rule diagrams proved to be the only practical key to the solution of hitherto insoluble problems concerning the mixing of components so that they would remain in equilibrium and not separate out and destroy the mixture. The phase rule helped make it possible to calculate in advance the temperature, pressure, and concentration required for stability—thus eliminating months and possibly years of tedious trial-and-error experiments. This would have important application in industry as well as in the laboratory.

Interestingly, after Gibbs's one major treatise on thermodynamics, he never wrote another important paper on the subject. He had said the last word, and he knew it. He did not, however, remain idle. Between 1883 and 1889, he published five papers on the electromagnetic theory of light. This work, too, was well received.

Meanwhile, he had begun to receive a certain amount of more or less perfunctory recognition at home: He was elected to the National Academy of Sciences in 1879 and to the American Academy of Arts and Sciences in 1880; in 1880 he received the Rumford Medal from the latter; in 1885 he was elected a vice president of the American Association for the Advancement of Science.

In the period between 1889 and 1902, Gibbs lectured on the subject of statistical mechanics but published almost nothing on the topic except for a brief abstract. This would be his major work during the final portion of his life; it would require about the same gestation period as did his investigation of thermodynamics. Simultaneously, however, he was lecturing and publishing papers on vector analysis and multiple algebra; the theory of dyadics that appeared in these works is regarded as his most important published contribution to pure mathematics. A book based on his lectures, *Gibbs' Vector Analysis*, was edited and published by a student, E. B. Wilson, in 1901.

During that same year, Gibbs was awarded the Copley Medal by the Royal Society of London for being the first to apply the second law of thermodynamics to the exhaustive discussion of the relation between chemical, electrical, and thermal energy and the capacity for external work. This was the highest honor for scientists prior to the founding of the Nobel Prize.

In 1902, Gibbs's final important work was published under the title *Elementary Principles in Statistical Mechanics Developed with Special Reference to the Rational Foundation of Thermodynamics*. In this brilliant study, Gibbs was as far ahead of his time as he had been with his first major treatise. The later work has been called "a monument in the history of physics which marks the separation between the nineteenth and twentieth centuries." Gibbs's perception of the role played by probability in physical events made his last work a true precursor to quantum mechanics, which did not develop fully until the 1920s.

During the year following the publication of his final gift to the world, Gibbs suffered from several minor ailments. One of these resulted in a sudden and acute attack from an intestinal obstruction, which led to Gibbs's untimely death on April 28, 1903.

Significance

Gibbs's contribution to American society occurred chiefly after his death. It is regrettable that very few Americans had the capacity to recognize his achievements while he was alive, but it seems pointless to try to fix the blame for this. On the one hand, he himself declined to make his papers more accessible by revising them for a wider readership. On the other, physical chemistry was only beginning to develop in the United States. Few professors of either chemistry or physics had the background that would have enabled them to understand Gibbs's work; there were no grand figures such as Rudolf Clausius, James Clerk Maxwell, William Thomson (Baron Kelvin), or Wilhelm Ostwald in the United States to welcome the new young genius personally.

In addition, the chemical industry in the United States was conservative in the matter of adopting new methods derived chiefly from theory, while at the same time it was caught up in the chaos of a greatly expanding industrialism. There was virtually no one available to examine the implications for the chemical industry of Gibbs's new phase rule. Gradually, however, as the industry turned more and more to synthesizing new compounds and developing metal alloys, there came a demand for precisely the sort of tool that Gibbs long before had provided. The phase rule had an early application in alloys of iron and carbon to produce different types of steel. Another application involved the industrial synthesis of ammonia from nitrogen and hydrogen, and of nitric acid from ammonia. Although most of these applications were first worked out in Europe, American industry finally learned how to reap the benefits of bringing theory to bear on practical processes. It finally came to recognize what it owed to Gibbs.

The United States thus reaped the practical benefits of Gibbs's work; it also had the honor of claiming as its own one of the world's greatest theoretical scientists.

Donald M. Fiene

Antoine-Laurent Lavoisier

French chemist

In addition to making important contributions to eighteenth century geology, physics, cartography, and economic reform (particularly in agriculture and manufacturing), Lavoisier, by discrediting the phlogiston theory and proving the law of the conservation of matter, founded modern chemistry. He also discovered and named gases, namely oxygen and nitrogen.

Areas of achievement: Chemistry, cartography, science and technology, business, economics

Born: August 26, 1743; Paris, France

Died: May 8, 1794; Paris, France

Early Life

Antoine-Laurent Lavoisier (ahn-twahn-law-rah lahv-wahz-yay) was born into an eminent bourgeois family. Over the preceding century, his family had risen gradually from humble origins. His father, who had inherited considerable wealth from an uncle as well as his position as attorney to the Paris parliament, was a lawyer. Émilie Punctis, Antoine's mother, was equally well positioned socially, being the daughter of a parliamentary advocate. The Lavoisiers lived in Paris, and Antoine remained in his parents' home until he married.

Lavoisier's schooling, which probably began in 1754, seems principally to have been in Paris's small but exclusive Collège Mazarin. Guided by tradition, he studied law, receiving his bachelor's degree in 1763 and his licentiate the following year. Not really engaged by his profession, Lavoisier began scientific studies with four of France's most eminent scientists: Abbé Nicolas-Louis de Lacaille, an astronomer and mathematician; Bernard de Jussieu, a botanist; Jean-Étienne

Guettard, a geologist and mineralogist; and Guillaume Rouelle, a chemist. Each of these men was a cynosure of Parisian intellectual circles who closely interacted with leading figures in Europe's scientific communities: individuals exceptional in an eighteenth century context for the breadth of their curiosities.

Three such superficially unrelated curiosities marked Lavoisier's earliest research. First, his inquiries into the properties of minerals, gypsum in particular, led to his invention in 1764 of plaster of Paris, a feat that produced his first published work by the Academy of Sciences. In 1765, he was encouraged by an academy-sponsored prize competition for development of the best night lighting for large towns. He won both the prize and a medal awarded by the king. Concomitantly, pursuing geological and cartological work with France's premier geologist, Lavoisier collaborated in the production of an official geological atlas of France. Reward was swift: Considered for nomination as an academician in 1766, he was elected in 1768, a vacancy having opened with the death of a renowned chemist, Théodore Baron. Remarkable vigor characterized his academy work, which ranged over hundreds of projects, such as the adulteration of cider, theories of color, and mesmerism.

Life's Work

Lavoisier's scientific milieu, as had been true for two millennia, essentially accepted that all natural phenomena were the result of various mutations of earth, air, fire, and water. The general belief was that these "elements" were inexplicable, hence the long-standing interest in alchemy—the transformation of something base into something precious or valuable to the prolongation of life, invariably accompanied by often indecipherable language explaining alchemists' experimentation. Some alchemical philosophizing had been eroded by several of Lavoisier's predecessors, such as Robert Boyle, and contemporaries such as Joseph Priestley. Boyle, for example, had demonstrated that combustion (fire) cannot occur in an airless vessel, nor could many other processes, including life. In the mid-1600s, however, German chemists, notably Johann Joachim Becher and Georg Ernst Stahl, contrived a theory that combustion required no air, only an "oily earth" dubbed phlogiston. Armed with this dictum, followers of the phlogiston theory could also explain numerous other chemical processes.

Anxieties about the purity of the Parisian water supply brought Lavoisier into conflict with such traditional assumptions and methods. No standards existed then to ascertain water's purity; indeed, common scientific opinions held that evaporating water was transformed into earth. Lavoisier, having reviewed the endeavors of his predecessors, went directly to the fundamentals of the water problem; namely, whether matter was transmutable. After a series of exquisite experiments, he demonstrated that the earth produced in distilling and evaporating water, in fact, came from abrasive actions of water on the glass. Thus, two thousand years of belief in the transmutability of matter was refuted.

Both Lavoisier's public responsibilities and his science were made less burdensome by his happy marriage in 1771 to Marie Paulze, a woman of wealth and important family connections, but, more significant, a person intellectually drawn to her husband's work. She not only translated scientific materials for him but also aided him in the niceties of his experiments and made excellent scientific sketches.

Air and fire preoccupied Lavoisier during the late 1760s and the early 1770s; that is, he was absorbed by the phlogiston question. During the 1760s, Priestley experimented with the properties of gases and discovered new ones, such as ammonia and sulfur dioxide. Heating the calx of mercury (a compound resulting from heating a substance below its melting point), Priestley produced a special "air," one in which candles burned vigorously and mice lived longer than in ordinary air. Yet Priestley, a believer in phlogiston, simply described this as dephlogisticated air.

Lavoisier launched an intricate series of experiments on combustion: Was the destruction of a diamond by heating merely an evaporation, or was it combustion? Was the same true of phosphorus, or of sulfur? By 1774, his answers appeared in his *Opuscules physiques et chymiques* (*Essays, Physical and Chemical*, 1776). Burning sulfur or phosphorus, instead of expelling something, actually absorbed air. Further reduction of the calx of lead unleashed air (now known as carbon dioxide) that sustained neither combustion nor life; in sum, reduction of the calx of metals expelled rather than absorbed something. By 1778 and 1779, continuing such investigations, Lavoisier reported that air consisted of two distinct gases: The first, which he named oxygen, sustained life and combustion and made up one-quarter of the volume, and the other, not respirable, made up the other three-quarters of the volume and is now identified as nitrogen. The phlogiston theory had been overthrown, and the foundations of modern chemistry emplaced, with his publication in 1786 of *Réflexions sur la phlogistique* (*Reflections on Phlogiston,* 1788). Lavoisier in 1787 then published his *Méthode de nomenclature chimique* (*Method of Chemical Nomenclature,* 1788), identifying a table of thirty-one chemical elements, adding heat and light as materials without mass. In full vigor, he proceeded conclusively to expound his oxygen theory, his new chemical nomenclature, and his statement of the law of the conservation of matter with publication in 1789 of a work rivaling any in the history of science: *Traité élémentaire de chimie* (*Elements of Chemistry, in a New Systematic Order, Containing All the Modern Discoveries*, 1790).

Genuine goodness and breadth of mind characterized both Lavoisier and his wife. Fine featured, delicate, and keenly intellectual, they look to each other lovingly, but with a hint of amused embarrassment, in a Jacques-Louis David portrait. Both an original experimenter and a superb scientific synthesizer, Lavoisier, aside from his establishment of modern chemistry, made immense contributions as a responsible, innovative public official. Though independently wealthy, Lavoisier had become a tax farmer to ensure his fortune. As a tax farmer, he relieved Jews from payment of the outrageous "cloven-hoof" tax; he sought relief for the poor from unjust taxation and instituted waste-cutting administrative reforms; he tried to protect honest Parisian merchants from smugglers who slipped by the *octrois* (a local import tax); he reformed the system and improved the quality of France's gunpowder production and generated a number of studies on the manufacture of saltpeter; like his brilliant successor, Alexis de Tocqueville, he sought the reformation of France's penal system; and his elaborate calculations of French agricultural conditions entitle him to front rank as a political economist. These are only partial indications of Lavoisier's genius beyond the bounds of chemistry.

As is so often the case, he died for his mistakes and for misinterpretations of his intentions. During the Reign of Terror during the French Revolution, he was imprisoned, charged with counterrevolutionary conspiracy against France, and executed in Paris on May 8, 1794. One French notable remarked to another in respect of Lavoisier: "Only a moment to cut off a head and a hundred years may not give us another like it." The remark, like Lavoisier's achievements, remains a tribute to eighteenth century French culture.

Significance

Antoine-Laurent Lavoisier, though a man of many talents that would have qualified him for eminence in half a dozen scientific disciplines or fields of public service, was the founder of modern chemistry not only in an abstract or theoretical dimension but also in immediately practical ways. His intellections and experimentation banished the phlogiston theory that had led and entranced many fine minds for centuries. In the tradition of Carolus Linnaeus, he sought only the facts, then classified them as they were verifiable through critical experimentation; he identified thirty-six chemical elements, not the least oxygen, and defined their characteristics and roles. He wrote the first precise texts, from which subsequent chemistry sprang, destined to be of incalculable benefit to humankind. Some of his work was entirely original; some was derivative; and, inevitably for great minds, some was magnificently synthetic. His official services to France ran a gamut that was encyclopedic and generally pragmatic, practical, and useful. He was preeminently the son of what then was Europe's greatest nation-state as well as its greatest center of mischief, intellection, and high culture.

Clifton K. Yearley

Dmitry Ivanovich Mendeleyev

Russian chemist

Although he did important theoretical work on the physical properties of fluids and practical work on the development of coal and oil resources, Mendeleyev is best known for his discovery of the periodic law, which states that the properties of the chemical elements vary with their atomic weights in a systematic way. His periodic table of the elements enabled him accurately to predict the properties of three unknown elements, whose later discovery confirmed the value of his system.

Area of Achievement: Chemistry

Born: February 8, 1834; Tobolsk (now Tyumen Oblast), Siberia, Russia

Died: February 2, 1907; St. Petersburg, Russia

Early Life

Dmitry Ivanovich Mendeleyev (myehn-dyih-LYAY-yehf) was born in an administrative center in western Siberia. He later recalled that he was the seventeenth child, but a sister claimed that he was actually the sixteenth child, and many scholars state that he was the fourteenth. His mother, Marya Dmitrievna Kornileva, came from an old merchant family with Mongolian blood. She became deeply attached to her youngest son and played an influential role in shaping his passionate temperament and directing his education. His father, Ivan Pavlovich Mendeleyev, was a principal and teacher at the Tobolsk high school, but shortly after his son's birth he became totally blind. The modest disability pension he received did not allow him to support his large family, and so Marya, a remarkably able and determined woman, reopened a glass factory that her family still owned in a village near

Tobolsk. She ran it so successfully that she was able to provide for her family and complete her younger children's education.

In the Tobolsk schools, young Dmitry, an attractive curly-haired, blue-eyed boy, excelled in mathematics, physics, geography, and history, but he did poorly in the compulsory classical languages, Latin in particular. Tobolsk was a place for political exiles, and one of Dmitry's sisters wedded an exiled Decembrist, one of those who tried unsuccessfully in December, 1825, to overthrow Czar Nicholas I. He took an active interest in Dmitry, taught him science, and helped form his political liberalism.

Toward the end of Mendeleyev's high school education, a double tragedy occurred: His father died of tuberculosis and his family's glassworks burned to the ground. By this time, the older children had left home, leaving only Dmitry and a sister with their mother, who decided to seek the help of her brother in Moscow. After Dmitry's graduation from high school in 1849, Marya, then fifty-seven years old, secured horses and bravely embarked with her two dependent children on the long journey from Siberia. In Moscow, her brother, after first welcoming them, refused to help his nephew obtain an education on the grounds that he himself had not had one. Marya angrily left Moscow for St. Petersburg, where she again encountered difficulty in getting her son either into the university or into the medical school.

Finally, through a friend of her dead husband, Mendeleyev's mother secured a place for him in the faculty of physics and mathematics of the Main Pedagogical Institute, the school his father had attended. Three months later, Marya died, and not long afterward her daughter succumbed to tuberculosis. Mendeleyev, who also suffered from tuberculosis, later wrote that his mother instructed him by example, corrected him with love, and, in order to consecrate him to science, left Siberia and spent her last energies and resources to put him on his way.

Mendeleyev received a good education at the Pedagogical Institute. One of his teachers had been a pupil of Justus von Liebig, one of the greatest chemists of the time. In 1855, Mendeleyev, now qualified as a teacher, was graduated from the Pedagogical Institute, winning a gold medal for his academic achievements.

Worn out by his labors, he went to a physician, who told him that he had only a short time to live. In an attempt to regain his health, he was sent, at his own request, to the Crimea, in southern Russia. He initially taught science at Simferopol, but when the Crimean War broke out, he left for Odessa, where he taught in a local lyceum during the 1855-1856 school year and where, aided by the warm climate, his health improved. In the autumn of 1856, he returned to St. Petersburg to defend his master's thesis. He succeeded and obtained the status of privatdocent, which gave him the license to teach theoretical and organic chemistry at the University of St. Petersburg.

Life's Work

In his teaching, Mendeleyev used the atomic weights of the elements to explain chemistry to his students. This did not mean that he believed that chemistry could be completely explained by physics, but his work on isomorphism and specific volumes convinced him that atomic weights could be useful in elucidating chemical properties. To improve his understanding of chemistry, he received in 1859 a stipend for two years' study abroad.

In Paris, Mendeleyev worked in the laboratory of Henry Regnault, famous for his studies on chlorine compounds, and at the University of Heidelberg, where he had the opportunity to meet Robert Bunsen, Gustav Kirchhoff, and other notable scientists. Because his weak lungs were bothered by the noxious fumes of sulfur compounds in Bunsen's laboratory, Mendeleyev set up a private laboratory to work on his doctoral thesis on the combination of alcohol and water. In the course of his research at Heidelberg, he discovered that for every liquid there existed a temperature above which it could no longer be condensed from the gas to the liquid form. He called this temperature the absolute boiling point (this phenomenon was rediscovered a decade later by the Irish scientist Thomas Andrews, who called it the "critical temperature," its modern descriptor). In September, 1860, Mendeleyev attended the Chemical Congress at Karlsruhe, Germany, and met the Italian chemist Stanislao Cannizzaro, whose insistence on the distinction between atomic and molecular weights and whose system of corrected atomic weights had a great influence on him.

Upon his return to St. Petersburg, Mendeleyev resumed his lectures on organic chemistry. Because he lacked a permanent academic position, he decided to write a textbook on organic chemistry, which became a popular as well as a critical success. In 1863, he began to act as a consultant for a small oil refinery in Baku. In this same year he was married to Fezova Nikitichna Leshcheva, largely because one of his sisters insisted that he needed a wife. The couple had two children, a boy, Vladimir, and a girl, Olga. The marriage was not happy, however, and quarrels were frequent. Eventually, Mendeleyev and his wife separated. He continued to live in their St. Petersburg quarters, while his wife and children lived at their country estate of Boblovo. In 1864, Mendeleyev agreed to serve as professor of chemistry at the St. Petersburg Technical Institute while continuing to teach at the university. A year later, he defended his doctoral thesis on alcohol and water, arguing that solutions are chemical compounds indistinguishable from other types of chemical combination.

A turning point in Mendeleyev's career occurred in October, 1865, when he was appointed to the chair of chemistry at the University of St. Petersburg. While teaching an inorganic-chemistry course there, he felt the need to bring to this subject the same degree of order that had characterized his earlier teaching of organic chemistry. Because he could find no suitable textbook, he decided to write his own. The composing of this book, eventually published as *Osnovy khimii* (1868-1871; *The Principles of Chemistry*, 1891), led him to formulate the periodic law. It was also one of the most unusual textbooks ever written. Unlike most textbooks, it was not a recycling of traditional material. It had instead a novel organization and an abundance of original ideas. It was also a curious blend of objective information and personal comment in which footnotes often took up more space than the text.

In organizing his ideas for the book, Mendeleyev prepared individual cards for all sixty-three elements then known, listing their atomic weights and properties, which showed great dissimilarities. For example, oxygen and chlorine were gases, whereas mercury and bromine were liquids. Platinum was very hard, whereas sodium was very soft. Some elements combined with one atom, others with two, and still others with three or four. In a search for order, Mendeleyev arranged the elements in a sequence of increasing atomic weights. By moving the cards around, he found that he could group certain elements together in already familiar families. For example, in the first table that he developed in March of 1869, lithium, sodium, potassium, and the other so-called alkali metals formed a horizontal row. In some groups he left empty spaces so that the next element would be in its proper family.

Mendeleyev's analysis of his first arrangement convinced him that there must be a functional relationship between the individual properties of the elements and their atomic weights. One of the many interesting relationships he noticed concerned valence, the ability of an element to combine in specific proportions with other elements. He observed a periodic rise and fall of valence—1, 2, 3, 4, 3, 2, 1—in several parts of his arrangement. Because valence and other properties of the elements exhibited periodic repetitions, he called his arrangement in 1869 the periodic table. At the same time he formulated the periodic law: Elements organized according to their atomic weights manifest a clear periodicity of properties. He had been thinking about information relevant to the periodic law for fifteen years, but he formulated it in a single day. Mendeleyev would spend the next three years perfecting it, and in important ways he would be concerned with its finer points until his death.

When Mendeleyev's paper was read by a friend at the Russian Chemical Society meeting in 1869, the periodic table did not evoke unusual interest. Its publication in German met with a cool reception. Mendeleyev's opponents, who were especially censorious in England and Germany, were suspicious of highly imaginative theoretical schemes of the elements; many scientists before Mendeleyev had proposed such systems, which resulted in little of practical benefit for chemists.

Mendeleyev believed that if he could convince scientists of the usefulness of his system, it would attract followers. Therefore, he tried to show how his table and periodic law could be used to correct erroneously determined atomic weights. More significant, he proposed in an 1871 paper that gaps in his table could be used to discover new elements. In particular, he predicted in great detail the physical and chemical properties of three elements, which he called eka-aluminum, eka-boron, and eka-silicon, after the Sanskrit word for "one" and the name of the element above the gap in the table.

Mendeleyev's predictions were met with great skepticism, but when, in 1875, Paul Lecoq de Boisbaudran discovered eka-aluminum in a zinc ore from the Pyrenees, skepticism declined, especially after chemists learned that the element's characteristics had been accurately foretold by Mendeleyev. When, in 1879, Lars Nilson isolated eka-boron from the ore euxenite, even fewer skeptics were to be found. Finally, in 1879, when Clemens Winkler in Germany found an element in the ore argyrodite that precisely matched Mendeleyev's predictions for eka-silicon, skepticism vanished. In fact, Winkler used Mendeleyev's predictions of a gray element with an atomic weight of about 72 and a density of 5.5 in his search (he found a grayish-white substance with an atomic weight of 72.3 and a density of 5.5). These new elements were given the names gallium (1875), scandium (1879), and germanium (1886), and their discovery led to the universal acceptance of Mendeleyev's periodic law.

In addition to campaigning for his periodic system, Mendeleyev during the 1870s spent time on his technological interests. He was a patriot who wanted to see such Russian resources as coal, oil, salt, and metals developed properly. With this in mind, he visited the United States in 1876 to study the Pennsylvania oil fields. He was critical of the American developers' concentration on the expansion of production while ignoring the scientific improvement of industrial efficiency and product quality. Upon his return to Russia, he was sent to study his country's oil fields, and he became critical of the way they were exploited by foreign companies. He urged Russian officials to develop native oil for the country's own benefit. From his experience in the oil fields, Mendeleyev developed a theory of the inorganic origin of petroleum and a belief in protective tariffs for natural resources.

In the year of his American trip, Mendeleyev underwent a domestic crisis. At a sister's home he had met Anna Ivanovna Popov, a seventeen-year-old art student, and fallen desperately in love with her. Anna's family opposed the relationship and made several attempts to separate the pair, resorting finally to sending her to Rome to continue her art studies. Mendeleyev soon followed her, leaving behind a message that if he could not wed her, he would commit suicide. She was mesmerized by this passionate man who, with his deep-set eyes and patriarchal beard, looked like a biblical prophet. She agreed to return to Russia and wed him, but the couple discovered that, according to the laws of the Russian Orthodox Church, Mendeleyev could not be remarried until seven years after his divorce.

Mendeleyev eventually found a priest who was willing to ignore the rule, but two days after the marriage the priest was dismissed and Mendeleyev was officially proclaimed a bigamist. Despite the religious crisis, nothing happened to Mendeleyev or his young wife. As the czar told a nobleman who complained about the situation: "Mendeleyev has two wives, yes, but I have only one Mendeleyev." The second marriage proved to be a happy one, and the couple had two sons and two daughters. Anna Ivanovna introduced her husband to art, and he became an accomplished critic and an astute collector of paintings.

During the 1880s and 1890s, Mendeleyev became increasingly involved in academic politics. Ultimately, conflict with the minister of education prompted him to resign from the University of St. Petersburg. At his last lecture at the university, where he had taught for more than thirty years, the students gave him an enthusiastic ovation. His teaching career at an end, Mendeleyev turned to public service, where he was active in many areas.

When the Russo-Japanese War broke out in February, 1904, Mendeleyev became a strong supporter of his country's efforts, and Russia's defeat disheartened him. By this time, Mendeleyev was not only the grand man of Russian chemistry but also, because of the triumph of the periodic law, a world figure. In 1906, he was considered for a Nobel Prize, but the chemistry committee's recommendation was defeated by a single vote, mainly because his discovery of the periodic law was more than thirty-five years old. Though he missed winning the Nobel Prize, he was showered with many awards in Russia and in many foreign countries. His end came early in 1907, when he caught a cold that developed into pneumonia. His chief consolation during his final illness was the reading of *A Journey to the North Pole* (*Les*

Anglais au pôle nord, 1864; English translation, 1874) by Jules Verne, his favorite author.

Significance

Dmitry Ivanovich Mendeleyev's name has become inextricably linked with the periodic table, but he was not the first to attempt to develop a systematic classification of the chemical elements. Earlier in the century, Johann Döbereiner, a German chemist, had arranged several elements into triads—for example, calcium, strontium, and barium—in which such properties as atomic weight, color, and reactivity seemed to form a predictable gradation. John Newlands, an English chemist, arranged the elements in the order of atomic weights in 1864 and found that properties seemed to repeat themselves after an interval of seven elements.

In 1866, Mendeleyev announced his "law of octaves," in which he saw an analogy between the grouping of elements and the musical octave. Several other attempts at a systematic arrangement of the elements were made before Mendeleyev, some of which were known to him. Many scholars credit the German chemist Lothar Meyer as an independent discoverer of the periodic law, since in 1864 he published a table of elements arranged horizontally so that analogous elements appeared under one another.

Other scholars contend that Mendeleyev's table was more firmly based on chemical properties than Meyer's and it could be generalized more easily. Furthermore, Mendeleyev was a much bolder theoretician than Meyer. For example, he proposed that some atomic weights must be incorrect because their measured weight caused them to be placed in the wrong group of the table (Meyer was reluctant to take this step). In most instances Mendeleyev's proposals proved to be correct (although the troublesome case of iodine and tellurium was not resolved until the discovery of isotopes). Finally and most notably, Mendeleyev was so impressed with the periodicity of the elements that he took the risk of predicting the chemical and physical properties of the unknown elements in the blank places of his table. Although his table had imperfections, it did bring similar elements together and help make chemistry a rational science and the periodic law an essential part of chemistry.

The periodic table grew out of the theoretical side of Mendeleyev's scientific personality, but he also had a practical side. He made important contributions to the Russian oil, coal, and sodium-carbonate industries. He served the czarist regime in several official positions. Nevertheless, he did not hesitate to speak out against the government's oppression of students, and his sympathy for the common people led him to travel third-class on trains. Though he held decidedly liberal views, it is wrong to see him as a political radical. Perhaps he is best described as a progressive, because he hoped that the czarist government would correct itself and evolve into a more compassionate regime.

Had Mendeleyev lived a few more years, he would have witnessed the complete and final development of his periodic table by Henry Moseley, whose discovery of atomic number by interacting X rays with various elements led to the use of the positive charge of the nucleus as the true measure of an element's place in the periodic table. Throughout the twentieth century, the periodic table, which owed so much to Mendeleyev, continued to be enlarged by the discovery of new elements. It was therefore appropriate that a new element (atomic number 101), discovered in 1955, was named mendelevium, in belated recognition of the importance of his periodic law.

Robert J. Paradowski

Linus Pauling

American chemist

Pauling won two unshared Nobel Prizes, and these prizes, in chemistry and in peace, symbolize his contributions. In the 1930s and 1940s his scientific discoveries helped to make the United States an important center for structural chemistry and molecular biology. In the 1950s and 1960s his activities in the peace movement helped to mobilize the American public against the atmospheric testing of nuclear weapons.

Areas of Achievement: Chemistry, biochemistry, biology, medicine, peace advocacy

Born: February 28, 1901; Portland, Oregon

Died: August 19, 1994; Big Sur, California

Early Life

Linus Pauling (LI-nuhs PAWL-ihng) was the son of Herman William Pauling and Lucy Isabelle (Darling) Pauling. His mother was a daughter of Oregon pioneers who could trace their ancestry in the United States to the seventeenth century. His father's family was German and had come to the United States after the revolutionary upheavals of 1848 in Europe. During childhood, Pauling and his two younger sisters led a peripatetic existence as their father, a traveling drug salesman, tried to find a position that suited him. The family eventually settled in Condon, Oregon, where, in a two-room schoolhouse, Linus's education began. Life in Condon proved to be financially unrewarding, however, and in 1909, the Paulings moved back to Portland. Shortly after settling in a new drugstore, Herman died suddenly of a perforated gastric ulcer. He was only thirty-two years old.

Herman's death created severe difficulties for the family. Pauling became a shy adolescent who spent most of his time reading. His intellectual energies also found an outlet in schoolwork, and he moved at an accelerated pace through Portland's individualized grammar school system and through Washington High School. The most important event of this period occurred when Lloyd Jeffress, a friend, showed him how sulfuric acid could turn white sugar into a steaming mass of black carbon. So excited was Pauling by what he saw that he decided then and there to become a chemist.

Pauling was able to pursue a career as a chemical engineer at Oregon Agricultural College by working at various jobs during the school year and in the summer. Because of the need to support his mother, he was forced to interrupt his education for a year and teach quantitative analysis. This hiatus gave him time to read science journals, and he came across papers on the chemical bond by Irving Langmuir and Gilbert Newton Lewis. These papers provoked his lifelong interest in chemical bonding and structure. In his senior year, he met Ava Helen Miller, a freshman in the general chemistry class he was teaching. At first she was not attracted to this curly-haired, blue-eyed young man who was "so full of himself," and Pauling, though attracted to her, was reluctant to show it because of his position as a teacher. After the course was over, and by the time of Pauling's graduation from college in 1922, they were very much in love.

Pauling's graduate career took place at the California Institute of Technology (Caltech) in Pasadena, where three professors, Arthur A. Noyes, Roscoe G. Dickinson, and Richard C. Tolman, helped to shape his career. Noyes acted as Pauling's father figure, and behind the scenes he made sure that his protégé remained a chemist (Pauling was being tempted by theoretical physics). Dickinson trained Pauling in X-ray diffraction, a technique for discovering the three-dimensional structures of crystals. Tolman was Pauling's mentor in theoretical physics. After a successful year at Caltech, Pauling returned to Oregon to marry Ava Miller. She returned with him to Pasadena, where they began a close relationship that continued until her death fifty-seven years later.

After receiving his Ph.D. from Caltech in 1925, Pauling spent a brief period as a National Research fellow in Pasadena. He was then awarded a Guggenheim Fellowship to study quantum mechanics in Europe. He spent most of his year and a half abroad at Arnold Sommerfeld's Institute for Theoretical Physics in Munich, but he also spent a month at Niels Bohr's institute in Copenhagen and a few months in Zurich. On his return to California in 1927, Pauling began a career as teacher and researcher at Caltech that would last for thirty-six years.

Life's Work

Structure was the central theme of Pauling's scientific work. Most of his early research focused on the determination of the structures of molecules, first by directing X-rays at crystals, later by directing electron beams at gas molecules. As these X-rays and electron techniques provided Pauling with experimental tools for discovering molecular structures, so quantum mechanics gave him a theoretical tool. For example, he used quantum mechanics to explain why the carbon atom forms equivalent bonds. In 1939, he wrote about many of his structural discoveries in *The Nature of the Chemical Bond and the Structure of Molecules and Crystals*, one of the most influential scientific books of the twentieth century.

Pauling's interest in biological molecules began in the 1930s with his studies of the hemoglobin molecule, whose striking red color and property of combining with oxygen appealed to him. Interest in hemoglobin led naturally to an interest in proteins, and with Alfred Mirsky he published a paper on the general theory of protein structure in which they suggested that proteins had coiled configurations that were stabilized by weak intermolecular forces and hydrogen bonds.

On one of Pauling's visits to the Rockefeller Institute to visit Mirsky, he met Karl Landsteiner, the discoverer of blood types, who introduced him to another field antibodies. Pauling's first paper on antibody structure appeared in 1940. During World War II, his work shifted toward practical problems, for example, the discovery of an artificial substitute for blood serum. This was only part of the extensive work that he did for the government. He also invented an oxygen detector, a device that depended on oxygen's special magnetic properties and that found wide use in airplanes and submarines. He also spent much time studying explosives and rocket propellants. At the end of the war, he became interested in sickle-cell anemia, which, he speculated, might be a molecular disease caused by an abnormal hemoglobin molecule. Working with Harvey Itano, Pauling showed in 1949 that this indeed was the case.

While a guest professor at Oxford University in 1948, Pauling returned to a problem that had occupied him in the late 1930s to find precisely how the chain of amino acids in proteins is coiled. By folding a piece of paper on which he had drawn such a chain, he discovered the alpha helix, a configuration of turns held together by hydrogen bonds, with each turn having a nonintegral number of amino-acid groups. Pauling and Robert B. Corey published a description of the helical structure of proteins in 1950, and this structure was soon verified experimentally.

During the early 1950s, Pauling became interested in deoxyribonucleic acid (DNA), and in February, 1953, he and Corey published a structure for DNA that contained three twisted, ropelike strands. Shortly thereafter, James D. Watson and Francis Crick published the double-helix structure, which turned out to be correct. Watson and Crick profited from X-ray photographs of DNA taken by Rosalind Franklin, photographs that were research tools denied Pauling because of the refusal of the U.S. State Department to grant him a passport for foreign travel. He was finally given a passport when he received the 1954 Nobel Prize in Chemistry for his research into the nature of the chemical bond.

With the heightened publicity given him by the Nobel Prize, Pauling began to devote more of his attention to humanitarian issues connected with science. For example, he became increasingly involved in the debate over nuclear fallout and in the movement against nuclear-bomb testing. In 1958, he and his wife presented a petition signed by more than eleven thousand scientists from around the world to Dag Hammarskjöld, secretary-general of the United Nations. Pauling defended his petition before a congressional subcommittee in 1960, and he risked going to jail by refusing to turn over his correspondence with those who helped to circulate the petition. For all these efforts, he was awarded the Nobel Peace Prize for 1962 on October 10, 1963, the day that the partial nuclear test-ban treaty went into effect.

Through the mid-1960s, Pauling was a staff member at the Center for the Study of Democratic Institutions (CSDI). He had left Caltech primarily because of the negative reaction of many members of the Caltech community to his peace efforts, and in Santa Barbara he hoped to be able to work effectively in both areas, science and peace. While at CSDI, he participated in discussions on peace and politics, and he proposed a new model of the atomic nucleus in which protons and neutrons were arranged in clusters.

Pauling left Santa Barbara in 1967 to become research professor of chemistry at the University of California, San Diego, where he published a paper on orthomolecular psychiatry that explained how mental health could be achieved by manipulating the concentrations of substances normally present in the body. During this time, Pauling's interest became centered on a particular molecule, ascorbic acid (vitamin C). He examined the published evidence about vitamin C and came to the conclusion that ascorbic acid, provided that it is taken in large enough quantities, helps the body fight off colds and other diseases. The eventual outcome of Pauling's work was the book *Vitamin C and the Common Cold*, published in 1970, while he was a professor at Stanford University.

This interest in vitamin C in particular and orthomolecular medicine in general led to his founding in 1973 the institute that now bears his name: the Linus Pauling Institute of Science and Medicine. During his tenure at the institute, Pauling was involved in a controversy about the relative benefits and dangers of the ingestion of large amounts of vitamins. In the early 1970s, he became interested in using vitamin C for the treatment of cancer, largely through his contact with Scottish physician Ewan Cameron. Their collaboration resulted in a book, *Cancer and Vitamin C* (1979), in which they marshaled evidence for the effectiveness of vitamin C against cancer. In the 1980s Pauling obtained financial support to have his ideas about vitamin C and cancer tested experimentally in his own laboratory as well as in the laboratories of the Mayo Clinic. Positive results were obtained in animal studies at the Linus Pauling Institute, but negative results were obtained with human cancer patients in two Mayo Clinic studies, and so the controversy remained unresolved.

Significance

Throughout his life Pauling saw himself as a Westerner, with a strong belief in those values that historian Frederick Jackson Turner called the "traits of the frontier": self-sufficiency, strength combined with inquisitiveness, a masterful grasp of material things, restless nervous energy, and love of nature and hard work. One can see these traits in Pauling's scientific work, as his curiosity drove him from one area to another. He liked to work on the frontiers of knowledge, not in crowded fields, and many of his greatest discoveries occurred in the area between disciplines between chemistry and physics, chemistry and biology, chemistry and medicine.

A master showman, Pauling cleverly fought for the recognition of his ideas. He was passionate in defending his scientific views, even when most of his colleagues did not accept them, as in his ideas about vitamin C. Despite his involvement in many scientific controversies, scientists have recognized the overwhelming importance of his contributions. On the occasion of Pauling's eighty-fifth birthday, Crick called him "the greatest chemist in the world." When Pauling was born, chemistry was a discipline dominated by Europeans, mainly Germans. Pauling's work symbolized and helped to make real the dominance of chemistry by Americans.

Pauling's influence on the United States was not restricted to chemistry. J. D. Bernal stated that if one person could be given credit for the foundation of molecular biology, that person would be Linus Pauling. Another field molecular medicine was created by Pauling's discovery of the first-known molecular disease, sickle-cell anemia. Besides helping revolutionize and found several scientific disciplines, Pauling had a major influence on American society through his many speeches and writings on peace. He helped to change the climate of American public opinion on nuclear weapons, which made the 1963 test-ban treaty possible. In fact, he was prouder of his efforts on behalf of peace than he was of his scientific accomplishments. Yet he did not see these contributions as separate: They were both part of his single-minded quest for truth.

Robert J. Paradowski

Robert Burns Woodward

American chemist

Woodward was the preeminent organic chemist in the postwar United States. Renowned for the total synthesis of complex natural products, for more than thirty years he achieved syntheses of unparalleled creativeness and elegance.

Area of Achievement: Chemistry

Born: April 10, 1917; Boston, Massachusetts

Died: July 8, 1979; Cambridge, Massachusetts

Early Life

Robert Burns Woodward was the only child born to Margaret Burns and Arthur Chester Woodward. His father died in October of 1918 at the age of thirty-three. His mother was a native of Glasgow, Scotland, and claimed descent from the poet Robert Burns. Young Woodward grew up in Quincy, near Boston, and attended its public schools. Chemistry attracted him from an early age, although he was vague on what started the interest. He indulged in this hobby in a home laboratory, fascinated from the start by the beauty of molecular structures. A gifted child, Woodward skipped grades three times, which enabled him to enroll at the Massachusetts Institute of Technology in 1933 at the age of sixteen. Possessing a staggering amount of chemical knowledge, he earned a Ph.D. from MIT within four years.

Woodward's experiences at MIT were not positive at first. The slow pace of the curriculum led to boredom and inattention to his studies, to such an extent that he was requested not to return for his sophomore year. Some of the faculty, however, recognized his unusual capabilities and arranged a reinstatement, which allowed him to forgo all requirements and class attendance as long as he took the course examinations. This unusual arrangement enabled him to devise his own program and take as many as fifteen courses in a semester. Furthermore, he requested laboratory space, then permitted only for graduate students; the faculty agreed to consider this if he would supply a list of experiments. His list was so original that he was given his own laboratory for the remainder of his stay at MIT. His doctoral dissertation was inspired by the lecture of a professor on the difficulties in synthesizing a sex hormone.

After a brief period in 1937 as instructor at the University of Illinois, Woodward returned to Boston, rising from Harvard University instructor to full professor in 1950, Morris Loeb Professor of Chemistry in 1953, and Donner Professor of Science in 1960, the latter position freeing him from lecturing.

Woodward was married twice, to Irja Pullman, whom he had known for many years, in 1938, and to Eudoxia Muller, a Polaroid consultant, in 1946; there were two children from each marriage. His passion for chemistry and his habit of pursuing it during all hours of the day and night made him difficult to live with, and both marriages ended in divorce.

Woodward was a tall, powerfully built man, very neat and well-groomed, his daily attire including a dark blue suit and a plain, light-blue necktie. He was addicted to puzzles and games and enjoyed being the life of a party, taking great pleasure as a storyteller. He developed a deep interest in literature of all types, especially biography. Woodward was an inveterate smoker and a heavy drinker of whiskey. From his youth onward, he normally slept only three hours a day. This ability to do without sleep, coupled with his incessant smoking and heavy consumption of alcohol, led him to believe that he was different from other people and would not be subject to the frailties of others. A lone wolf throughout his life, he made few intimate friends and, despite appearances, he was a very lonely human being.

Life's Work

When Woodward began his research at Harvard, synthetic organic chemistry had major limitations. Chemists were still using the same techniques of separation and purification of natural products as had chemists of 1890. To obtain a total synthesis of a natural product was an enormous challenge, especially with the lack of control over the essential three-dimensional arrangement of atoms in space (stereochemistry). Yet Woodward chose to specialize in this realm of naturally occurring animal and plant substances because it was one of endless fascination and unlimited opportunities.

Woodward's success lay in combining two contemporary developments in chemistry. An instrumental revolution was taking place during his research career, with ultraviolet, infrared, and nuclear magnetic resonance spectroscopy coming into general use by chemists as well as novel methods of chromatography for the separation and purification of substances. He combined these and other tools of the instrumental revolution with a master plan based on theoretical developments involving electronic reaction mechanisms. He could determine theoretically how chemical bonds form, break, and rearrange, and the stereochemical configurations that would result, and envisage the entire synthetic route before any laboratory work was done. In this way, Woodward made organic synthesis into both an art of creative planning and design and an experimental science involving hard, tedious laboratory work. A typical Woodward synthesis was noted for its planning. In the execution of the plan, he rarely took a physical part. His reputation assured him of coworkers of high quality; his role was to direct and inspire them.

The revolution in organic synthesis was first announced in 1944 when Woodward, along with another brilliant young chemist, William Doering, synthesized the antimalarial drug quinine. It was important enough to make page one of *The New York Times*, since it was part of the World War II effort to find quinine substitutes. (Natural quinine came from cinchona tree bark, and the plantations were in Japanese control.) Woodward and Doering achieved the synthesis in fourteen months, starting with a common coal-tar derivative, and showed what could be done with an original plan, the mapping of every step, and resourceful experimental work.

Following the quinine synthesis, Woodward turned to another important wartime project: penicillin, then a strategic material desperately needed to combat infections. In 1945, he was the first to propose, defend, and show the consequences of the fused ring structure basic to all penicillins. He continued to make important contributions to antibiotic research over the years. He investigated the broad spectrum antibiotics terramycin and aureomycin following their discovery, establishing the complete structures of these tetracyclic molecules by 1952.

After the war, Woodward continued to pursue problems of structure determination and synthesis. In 1947, he solved the structure of the complex natural poison strychnine in a brilliant interpretation of an array of experimental data and transformations. He also synthesized strychnine in 1954; it was one of several important syntheses of alkaloid drugs.

Woodward made headlines again in 1951 with the first total synthesis of a natural steroid. After sixteen months of hard work, involving twenty steps from a coal-tar derivative, he obtained the steroid nucleus, which, since many steroids had already been interconverted, represented a synthesis of many of them, such as cortisone and cholesterol. *Time* and *Newsweek* featured stories and interviews with him. Cortisone, made with difficulty from oxbile, was scarce and expensive. His synthesis was from relatively cheap coal tar and promised an abundance of the antiarthritis drug. Woodward tried to dampen the enthusiasm; his was only a laboratory synthesis. In fact, his feat was practical, and cortisone came to be provided to arthritis sufferers in synthetic form.

While most of his synthetic work concerned natural products, Woodward made important contributions to other areas. In 1952, in one of his most inspired insights, he revealed the structure of dicyclopentadienyl iron or "ferrocene," as he named it. Discovered in 1951, its nature was intensely debated, and Woodward daringly proposed the famed sandwich structure of an atom of iron sandwiched between two cyclopentadienyl rings. This was a new type of organometallic substance and the basis of a new class of molecules having unique pi electron binding. (Pi bonding refers to modern molecular orbital theory in organic chemistry.)

One brilliant synthesis after another marked the 1950s. Woodward followed the 1954 synthesis of lysergic acid, the basis for the hallucinogenic drug LSD, with a masterful synthesis in 1956 of reserpine, the first of the tranquilizing drugs. The reserpine synthesis stood out for its complete stereospecificity and high yield; it was the most elegant synthesis ever achieved and, as with cortisone, so well done that his plan could be adapted to commercial production.

Woodward kept reaching for higher levels of complexity and took on the challenge of chlorophyll, the green pigment in plants responsible for the conversion of solar energy into the basic materials of life. The synthesis took four years (1956-1960) of work by his dedicated research group of Harvard graduate students and postdoctoral fellows.

The year 1965 was noteworthy for two developments. The first was Woodward winning the Nobel Prize "for his outstanding achievement in the art of organic synthesis." The wording is significant; the honor was not only for the synthesis of natural products but also for the distinctive way that the synthesis was achieved. Interestingly, in his address on receiving the prize, Woodward did not reflect on his past accomplishments but used the occasion to illuminate the "art" of synthesis by reporting on his latest work, the synthesis of the antibiotic cephalosporin C.

The cephalosporin work was not done at Harvard, for in 1963, Woodward acquired a second research group with the creation of the Woodward Research Institute in Basel, Switzerland, by Ciba Limited, with Woodward as director; like Harvard, it became a center of impressive achievements.

Also in 1965 came what may have been Woodward's most significant contribution to organic chemistry. He was deeply interested in theoretical chemistry and had published in this area from time to time. With a distinguished collaborator, Roald Hoffmann (a Cornell professor and Nobel Prize winner in 1981), Woodward disclosed a new principle, the conservation of orbital symmetry. This discovery has been quoted more frequently than anything else he published. The conception came out of his synthetic work in which some puzzling stereochemical phenomena appeared. Baffled, he and Hoffmann eventually realized that something fundamental was behind the unexpected results. The explanation was the conservation of orbital symmetry, a quantum mechanical description of chemical bonding and reactivity, extending it to the course and stereochemistry of organic reactions and allowing for the prediction of both the way a reaction would proceed and the stereochemistry of the products. This principle stands as one of the most basic theoretical advances in the history of organic chemistry and almost immediately changed profoundly the thinking of chemists.

The most formidable problem undertaken by Woodward was the total synthesis of vitamin B_{12}; its complexity was greater than anything he had ever encountered. With another collaborator, Albert Eschenmoser of Zurich, and about one hundred coworkers, Woodward devised a strategy involving a pyrotechnic display of appearing and disappearing rings with the vitamin appearing at the end of an eleven-year hunt in 1972.

Woodward never stopped working; there were always more challenges. From his Basel group came the synthesis of a prostaglandin in 1973, a member of the relatively new and fascinating class of biologically active substances. He even joined with Hoffman after a ten-year interval in an attempt to design some molecular systems with novel conducting properties, a new field to both men and testimony to his versatility. At the time of his death following a heart attack at his Cambridge home, he was planning the synthesis of another antibiotic.

In addition to the achievements that appeared in his publications, Woodward's lectures and seminars were a major source of influence. He held lectureships worldwide, and few scientists could match the elegance, precision, and great artistic quality of his talks at the blackboard, with formulas and reaction sequences outlined in a unique graphic style that produced both clarity and aesthetic satisfaction. Every presentation was dramatic, creative, and lucid. Woodward enjoyed the attention that he commanded, relishing holding an audience enthralled for hours.

By the end of his life, Woodward had become the most honored of American scientists. In addition to the Nobel Prize, there was the National Medal of Science presented by President Johnson in 1964 and some thirty international awards, including the Pius XI Gold Medal of the Pontifical Academy of Sciences (1961) and the Order of the Rising Sun from Japan (1970). Some twenty-five universities in North America, Europe, Asia, and Australia awarded him honorary degrees, and almost all major scientific societies made him an honorary member. In 1979, in the announcement of his death released by Harvard, Woodward was described as "the greatest organic chemist of modern times."

Significance

Woodward's mastery of all available physical, instrumental, and theoretical methods guided him through the pathways of organic reactions and his astonishing array of achievements. He combined the three major strands in modern organic chemistry, the structural and synthetic, the physical and theoretical, and the instrumental, and set new standards of quality with his syntheses and structural elucidations of complex natural products. He taught others how to think and plan in detail; he taught theoreticians how to interact with experimentalists in the quest for solutions. Woodward represented a new generation of organic chemists who combined the best of modern theoretical and experimental methods in highly original and creative ways to solve incredibly complex and difficult problems.

Albert B. Costa

Bibliography

"A Noble Soul." *Shimadzu.com.* Shimadzu Corporation, n.d. Web. 9 Aug. 2012.

"All That Pollutes Isn't Bad." *McGill Reporter* 33.15. McGill University, 19 Apr. 2001. Web. 9 Aug. 2012.

"Carolyn R. Bertozzi." *Bertozzi Research Group: University of California, Berkeley,* n.d. Web. 13 August 2012.

"Commander Peggy Whitson Breaks Record for Time in Space for a U.S. Astronaut." *ScienceDaily.* ScienceDaily, 21 Apr. 2008. Web. 9 Aug. 2012.

"Dr. Don Catlin Responds to SI Article on Lance Armstrong." *Outside: Adventure,* 27 January 2011. Web. August 13 2012.

"Eureka for NERICA!" *New Agriculturist.* WRENmedia, July 2004. Web. 13 Aug. 2012.

"Global Warming's Big Thinkers: Angela Belcher." *Time Specials.* Time Inc., n.d. Web. 13 August 2012.

"Hungarian-born Nobel Prize Winner to visit Hungary." Europe Intelligence Wire. 8 Oct. 2004. Print.

"Jennifer A. Doudna, Ph.D." *HHMI Investigators.* Howard Hughes Medical Institute. Web. 13 August, 2012.

"Jennifer A. Doudna." *Molecular and Cell Biology.* University of California at Berkeley. Web. 13 August 2012.

"JGI Director Eddy Rubin." *DOE Joint Genome Institute.* The Regents of the University of California. Web. 9 Aug. 2012.

"Joan Argetsinger Steitz." *ASCB Newsletter* Jun 2006: 18–20. Print.

"Koichi Tanaka." *Nobelprize.org.* Nobel Media, n.d. Web. 9 Aug. 2012.

"Mahabir P. Gupta." *AAAS Awards: 2003 Award Recipients.* AAS. n.d. Web. 13 August 2012.

"Mary Good." *The Heinz Awards.* Heinzawards.net. 1 August 2004. Web. 13 August 2012.

"Mary Lowe Good, PhD." *The Catalyst Series: Women in History.* Chemical Heritage Foundation. n.d. Web. 13 August 2012.

"Milestones in Don Catlin's Career." *USA Today: Sports,* 28 February 2007: C9. Print.

"Neanderthal Genome Sequencing Yields Surprising Results and Opens a New Door to Future Studies." *Research News.* Berkeley Lab, 15 Nov. 2006. Web. 9 Aug. 2012.

"Profile: Aaron Ciechanover." *Agence France-Presse* 1 January 2001. Print.

"Scientist Wins Prize for New African Rice." *Africa Renewal* 18.4 (Jan. 2005): 9. Print.

"Spencer Wells: 'At Root, We're Still Hunters.'" *Independent.co.uk*. The Independent, 7 June 2010. Web. 9 Aug. 2012.

"Tree Bark Could Fight AIDs." *BBC News*. 20 February 2000. Web. 13 August 2012.

"U.S. Meeting Acknowledges Tribal Healers." *Irish Times.com*, 2 February 2000. Web. 13 August 2012.

"UCLA Drug Lab Faces Loss of IOC Accreditation." *Los Angeles Times* 4 December 1997, n.d. Print.

Angier, Natalie. "Scientist at Work: Jacqueline Barton." *New York Times* 2 March 2004: F1. Print.

Barth, Amy. "Scientist of the Arab Spring." *Discover Magazine*. Kalmbach, 3 Jan. 2012. Web. 9 Aug. 2012.

Barton, Jacqueline K. "Coping with technological protectionism." *Harvard Business Review* 6 (November/December 1984). Print.

Belcher, Angela. "Biologically Templated Photocatalytic Nanostructures for Sustained Light-driven Water Oxidation." *Nature Nanotechnology* 5 (2012): 340–44, 2010. Print.

---.. "Fabricating Genetically Engineered High-Power Lithium Ion Batteries Using Multiple Virus Genes." *Science* 32 (2009): 1051. Print.

---. "Virus-templated Self-assembled Single-walled Carbon Nanotubes for Highly Efficient Electron Collection in Photovoltaic Devices." *Nature Nanotechnology* 6 (2011): 377–84. Print.

Ben-Ari, Elia T. "Molecular Biographies." *Bioscience* 49.2 (Feb. 1999): 98. Print.

Berry, K. E., Waghray, S., Doudna, J. A. "The HCV IRES pseudoknot positions the initiation codon on the 40S ribosomal subunit." *RNA* 16.8 (2010): 1559–1569. Print.

Block, Melissa. "Discovering the Genetic Controls That Dictate Life Span" *Life Extension Magazine* June 2002: 51. Print.

Brown, David. "Israeli, American Chemists Win Nobel." *Washington Post* 7 Oct. 2004: A3. Print.

Brown, Eryn. "Building Chips. One Molecule at a Time." *Fortune* 14 May 2001: 166. Print.

Bryant, Christa Case. "Gatekeeper for Clean Sports." *Christian Science Monitor* 5 (August 2008): 25. Print.

Caplan, Jeremy. "The TIME 100: Kari Stefansson." *Time*. Time, 3 May 2007. Web. 9 Aug. 2012.

Carey, Elaine. "Female Scientist A Hero in Her Field: Yale's Joan Steitz, 65 Honoured." *Toronto Star* 3 Apr. 2006: A4. Print.

Carreau, Mark. "Astronaut Chronicles Her 5-Month Voyage." *Houston Chronicle* 10 Nov. 2002: A1. Print.

Catlin, Don. Interview by Peter Aldhous. "Confronting Drugs in Sports." *New Scientist* 8 August 2007: 2616. Print.

Bibliography

Chang, Kenneth. "3 Whose Work Speeded Drugs Win Nobel." *New York Times* 10 Oct. 2002: A33. Print.

Chizamdzhev, Yuri A., Kuzmin, Peter I, Weaver, James C. and Russel O Potts. "Skin Appendageal Macropores as Possible Pathway for Electrical Current." *Journal of Investigative Dermatology Symposium Proceedings* 3 (1998): 148–152. Print.

Ciechanover, Aaron, Hod, Y. and Hershko, A. (1978). "A Heat-stable Polypeptide Component of an ATP-dependent Proteolytic System from Reticulocytes." *Biochem. Biophys. Res. Commun.* 81 (1978): 1100–1105. Print.

---. "Proteolysis: from the lysosome to ubiquitin and the proteasome." *Nat. Rev. Mol. Cell Biol.* 6 (January 2005): 79–87. Print.

Cole, K. C. "Putting DNA to Work as a Biomedical Tool." *Los Angeles Times* 29 Dec. 1997. Web. 12 Aug.

Cook, Lynn J. "Millions Served." *Forbes*, 23 Dec. 2002: 302. Print.

Corcoran, Elizabeth. "The Next Small Thing." *Forbes* 23 July 2001. Print.

Djerassi, Carl. This Man's Pill: Reflections on the 50th Birthday of the Pill (Popular Science). New York: Oxford UP, 2004. Print.

---. *Cantor's Dilemma*. New York: Doubleday, 1989. Print.

---. *The Pill, Pygmy Chimps, and Degas' Horse: Autobiography of Carl Djerassi*. New York: Basic, 1992. Print.

Dreifus, Claudia. "A Conversation With Eva Harris; How the Simple Side of High-Tech Makes the Developing World Better." *New York Times* 30 Sept. 2003: F3. Print.

---. "The Body's Protein Cleaning Machine." *New York Times.* 19 June 2012: D2. Print.

Dunn, Ashley. "Molecular Computer Takes a Major Stride." *Los Angeles Times.* Los Angeles Times, 16 July 1999. Web. 9 Aug. 2012.

Evangelista, Andy. "Cynthia Kenyon: Probing the Prospects of Perpetual Youth." *UCSF Magazine.* The Regents of the University of California. May 2003, Web. 9 Aug. 2012.

Fox, Barry. "Frozen Stiff." *New Scientist* 2145 (1 Aug. 1998). Print.

Gedik, Nuh. "Science in the Islamic World: An Interview With Nobel Laureate Ahmed Zewail." *The Fountain* 67 (January 2009). Print.

Good, Mary L. Interview by Traynham, James G. *Mary L. Good.* Philadelphia, PA: Chemical Heritage Foundation, 2 June 1998. Print.

Grafov, Damaskin, and Pleskov Krishtalik. "Chronicle: Yuri Aleksandrovich Chizmadzhev." *Russian Journal of Electrochemistry.* 37.12 (2001): 1320–1321. Print.

Gupta, Mahabir P., Handa S.S., and K. Vasisht."Biological screening of plant constituents, v ICS/UNIDO."*International Centre for Science and High Technology, Trieste* (2006): 114. Print.

Hall, Carl T. "How Sleuths Tracked a Mysterious Steriod." *San Francisco Chronicle* 26 October 2003. Print.

Hershko, A., and Aaron Ciechanover. "Mechanisms of intracellular protein breakdown." *Annu. Rev. Biochem.* 51 (1982): 335–364. Print.

Hively, Will. "Married to the Molecule." *Discover* 1 October 1994. Kalmbach P. Print.

Houlton, Sarah. "Company Profile: Chemical Ensemble."*Chemistry World* Mar. 2011. Print.

Ioset, J.R., Maston, Andrew, Gupta, Mahabir, P., and Kurt Hostettmann. "Antifungal Xanathones from Roots of Marila Laxiflora." *Pharmaceutical Biology* 36.2 (1998): 103–106. Print.

Jacqueline K. Barton. Chemical Heritage Foundation, Fall 2002. Web. 13 August 2012.

Jinek, M., Coyle, S. M., Doudna, J. D."Coupled 5' nucleotide recognition and processivity in Xrn1-mediated mRNA decay." *Mol. Cell* 41.5(2011): 600–8. Print.

Johnson, Mark. "Persistence, Genius Mix for Chemist." *JSOnline.* Milwaukee Journal Sentinel, 30 Dec. 2007. Web. 9 Aug. 2012.

Kenny, Ursula. "Drug Innovators: Dr. Nick Terret and Dr. Ian Osterloh, Viagra." *The Observer.* Guardian News and Media, 30 Mar. 2002. Web. 9 Aug. 2012.

Kilian, Michael, and Jeremy Manier. "Your Ancestry Might Surprise You." *Chicago Tribune* 14 Apr. 2005: C12. Print.

Kolbert, Elizabeth. "Sleeping With the Enemy." *The New Yorker.* Condé Nast, 15 Aug. 2011. Web. 9 Aug. 2012.

Kotler, Steven. "Basic Research: Cynthia Kenyon." *Discover Magazine* Nov. 2004. Print.

Kreisler, Harry. "Conversation with Eva Harris." *Conversations with History.* Regents of the University of California, 15 Mar. 2001. Web. 9 Aug. 2012.

Long, Janice R. "Mary Lower Good to Receive 1997 Priestley Medal." *Chemical and Engineering News,* May 13, 1996. Web. 13 August 2012.

Lruse, Loren. "Hope for a New Direction." *Successful Farming,* 1 Nov. 2005. Print.

Mabunda, Doris. "Florence Wambugu." *Contemporary Black Biography.* The Gale Group, 9 Aug. 2012.

Mader, Becca. "Shaw Scientist Grants Provide Support To Promising Researchers." *The Business Journal* 26 Apr. 2002: 21. Print.

Marino, Melissa. "Biography of Jennifer A Doudna." *Proceedings of the National Academy of Sciences* (7 December 2004): 16987. Print.

Marris, Emma. "Straight Talk From...Eva Harris." *Nature Medicine.* Nature Publishing Group, 2007. Web. 9 Aug. 2012.

Maugh, Thomas H. II. "3 Win Nobels in Chemistry for Molecular Studies." *Los Angeles Times* 10 Oct. 2002: I31. Print.

---. "Ovary Removal May Cut Cancer Risk." *Los Angeles Times.* Los Angeles Times, 21 May 2002. Web. 9 Aug. 2012.

Bibliography

Munson, Kyle. "America's Uncertain Mission." *Des Moines Register*. Gannett, 28 Jan. 2011. Web. 9 Aug. 2012.

O'Keefe, Michael. "Scientist Don Catlin Understands NFLPA's Hesitation to Allow WADA to Test Players for HGH." *NY Daily News* 29 August 2011: 62. Print.

Olson, Steve "Neanderthal Man." *Smithsonian.com*. Smithsonian Media, Oct. 2006. Web. 9 Aug. 2012.

Pachauri, Rajendra. "The 2008 TIME 100: Susan Solomon." *TIME Magazine*. Time, 30 Apr. 2009. Web. 10 Aug. 2012.

Peterson, Kristin. "Dr. Olufunmilayo Olopade to Present Commencement Speech at Dominican University." *TribLocal Oak Park/River Forest*. Tribune Company, 20 Dec. 2010. Web. 9 Aug. 2012.

Preussl, Paul."Cellular Engineers Design Custom cell Surfaces Able to adhere to Synthetic materials." *Berkeley Lab: Research News* 5 (January 1998). Print.

Revkin, Andrew C. "Melding Science and Diplomacy to Run a Global Climate Review." *New York Times*. New York Times Company, 6 Feb. 2007. Web. 9 Aug. 2012.

Rincon, Paul. "Three Share Chemistry Novel Prize." *BBC News*, 5 October 2005. Web. 13 August 2012. Print. Web.

Rolnik, Guy, and Oren Majar. "A meeting of Minds on Matter, and Spirit." *Haaretz* 19 July 2011: AV 22. Print.

Rouhi, Maureen A. "Olefin Metathesis: The Early Days." *Chemical & Engineering News*. 23 (December 2002): 34–38. Print.

Roy, Karen. "A Different Tune." *Chicago Tribune* 26 Oct. 1997: WomanNews 3. Print.

Sachs, Jeffrey. "The TIME 100: Monty Jones." *TIME Magazine*. Time, 3 May 2007. Web. 13 Aug. 2012.

Sakai, Jill. "Two Faculty Elected to National Academy of Sciences." *University of Wisconsin-Madison News*. Board of Regents of the University of Wisconsin System, 1 May 2007. Web. 9 Aug. 2012.

Sample, Ian. "Rare Genetic Mutation Gives Hope In Battle Against Alzheimer's." *The Guardian*. Guardian News and Media, 11 July 2012. Web. 9 Aug. 2012.

---. "The Giants of Science." *The Guardian* 25 Sept. 2003: 4. Print.

Sashital, D. G., Jinek, M., Doudna, J. A. "An RNA-induced conformational change required for CRISPR RNA cleavage by the endoribonuclease Cse3." *Nat. Struct. Mol. Biol.* 18.6 (2011): 680–7. Print.

Schuman, Tomas. *Love Letter to America*. Los Angeles: NATA, 1984. Print.

Shallwani, Pervaiz. "Bay Area Surprise Winners: UC Berkely Hits the Jackpot with Nation's Best crop of MacArthur Fellowship Recipients." *San Francisco Chronicle* 23 June 1999. Web. 13 August 2012.

Shaughnessy, Dan. "Lab is helping baseball to clean up its act." *The Boston Globe* 16 July 2006: C p1. Print.

Shute, Nancy. "The Human Factor." *USNews.com*. U.S. News & World Report, 20 Jan. 2003. Web. 9 Aug. 2012.

Stein, Rob. "American, French Chemist Win Novel." *Washington Post* 6 October 2005: A12. Print.

Swaminathan, Nikhil. "Genomic 'Time Machine' May Pinpoint Divergence of Human and Neandertal." *Scientific American*. Nature America, 15 Nov. 2006. Web. 9 Aug. 2012.

Sweet, Lynn. "Another Chicagoan, Olufunmilayo Falusi Olopade, Gets Obama White House Post." *Chicago Sun-Times*. Sun-Times Media, 25 Feb. 2011. Web. 9 Aug. 2012.

Taubes, Gary. "Speed Demon." *UCLA Magazine*. The Regents of the University of California, Spring 2000. Web. 9 Aug. 2012.

Wade, Nicholas. "Gene May Play a Role in Schizophrenia." *New York Times*. New York Times Company, 13 Dec. 2002. Web. 9 Aug. 2012.

---. "Still Evolving, Human Genes Tell New Story." *New York Times* 7 Mar. 2006: A1. Print.

Walker, Gabrielle. "In From the Cold." *New Scientist* 13 Oct. 2001: 4646. Print.

Walsh, Mary Williams. "A Big Fish in a Small Gene Pool." *Los Angeles Times* 5 June 1998: A1. Print.

Wambugu, Florence. "Protesters Don't Grasp Africa's Need." *Los Angeles Times* 11 Nov. 2001. M1. Print.

Wang, Linda. "Carolyn Bertozzi Named Kavli Lecturer." *Chemical & Engineering News* 89.31 (1 August 2011): 48. Print.

Williams, Nigel. "NO New Is Good News-But Only for Three Americans." *Science* 23 Oct. 1998: 610–611. Print.

Williams, Zoe. "How the Inventor of the Pill Changed the World for Women." *Guardian*, 29 October 2012. Print.

Woodbury, Margaret A. "Trailblazer Turned Superstar." *Howard Hughes Medical Institute Bulletin* 19.1 (Feb. 2006): 21. Print.

Wüthrich, Kurt. "Autobiography." *Nobelprize.org*. Nobel Media, 2002. Web. 10 Aug. 2012.

Zewail, Ahmed. "Autobiography." *Nobelprize.org*. Nobel Media, Jan. 2006. Web. 9 Aug. 2012.

Selected Works

Bertozzi, Carolyn, and Peng George Wang, eds. *Glycochemistry: Principles, Synthesis, and Applications*. New York: Dekker, 2001. Print

Djerassi, Carl. *Cantor's Dilemma*. New York: Penguin, 1989. Print.

Djerassi, Carl. *The Pill, Pygmy Chimps, and Degas' Horse: The Autobiography of the Carl Djerassi*. New York: Basic, 1992. Print.

Djerassi, Carl. *From the Lab into the World: A Pill for People, Pets, and Bugs*. Washington, DC: American Chemical, 1994. Print.

Djerassi, Carl. *This Man's Pill—Reflections on the 50th Birthday of the Pill*. 2001

Selected Plays: *An Immaculate Misconception*, 1998; *Oxygen*, 2001. Print.

Catlin, Don H., and Thomas H. Murray. "TH Performance-enhancing drugs, fair competition, and Olympic sport." *Journal of the American Medical Association* 276.3 (1996): 231–37. Print.

Harris, Eva. *A Low-Cost Approach to PCR: Appropriate Transfer of Biomolecular Techniques*. New York: Oxford UP, 1998. Print.

Solomon, Susan. *The Coldest March: Scott's Fatal Antarctic Expedition*. Chicago: Donnely, 2001. Print.

Wambugu, Florence. *Modifying Africa: How Biotechnology Can Benefit the Poor and Hungry, a Case Study From Kenya*. Kenya: Wambuga, 2001. Print.

Wells, Spencer. *The Journey of Man: A Genetic Odyssey*. Princeton: Princeton UP, 2002. Print.

---. *Deep Ancestry: Inside the Genographic Project*. Washington D.C.: National Geographic, 2006. Print.

---. *The Journey of Man: A Genetic Odyssey*. National Geographic, 2003. Film.

---. *Explorer: Quest for the Phoenicians*. National Geographic, 2004. Film.

---. *The Search for Adam*. National Geographic, 2005. Film.

---. *China's Secret Mummies*. National Geographic, 2007. Print.

Indexes

Geographical Index

Austria
    Djerassi, Carl, 56

Connecticut
    Catlin, Don H., 27

Egypt
    Zewail, Ahmed H., 218
England
    Heath, James R., 86
    Terrett, Nicholas, 187

France
    Chauvin, Yves, 37

Georgia
    Wells, Spencer, 199

Hawaii
    Doudna, Jennifer, 63
Honduras
    Moncada, Salvador, 122
Hungary
    Hershko, Avram, 94

Iceland
    Stefansson, 164
Illinois
    Kenyon, Cynthia, 107
    Solomon, Susan, 157
India
    Gupta, Mahabir P., 74
Iowa
    Whitson, Peggy, 207
Israel
    Ciechanover, Aaron, 47

Japan
    Tanaka, Koichi, 182

Kenya
    Wambugu, Florence, 192

Massachusetts
    Bertozzi, Carolyn R., 21
Minnesota
    Steitz, Joan A., 174

New York
    Barton, Jacqueline K., 3
    Harris, Eva, 78
    Rubin, Edward M., 146
Nigeria
    Olopade, Olufunmilayo, 130

Russia
    Chizmadzhev, Yuri Aleksandrovich, 43

Sierra Leone
    Jones, Monty, 101
Sweden
    Paabo, Svante, 138
Switzerland
    Wüthrich, Kurt, 212

Texas
    Belcher, Angela, 12
    Good, Mary L., 69

Wisonsin
    Kiessling, Laura, 118

Name Index

Barton, Jacqueline K., 3
Belcher, Angela, 12
Bertozzi, Carolyn R., 21

Catlin, Don H., 27
Chauvin, Yves, 37
Chizmadzhev, Yuri Aleksandrovich, 43
Ciechanover, Aaron, 47

Djerassi, Carl, 56
Doudna, Jennifer, 63

Good, Mary L., 69
Gupta, Mahabir P., 74

Harris, Eva, 78
Heath, James R., 86
Hershko, Avram, 94

Jones, Monty, 101

Kenyon, Cynthia, 107
Kiessling, Laura L., 118

Moncada, Salvador, 122

Olopade, Olufunmilayo, 130

Paabo, Svante, 138

Rubin, Edward M., 146

Solomon, Susan, 157
Stefansson, Kari, 164
Steitz, Joan A., 174

Tanaka, Koichi, 182
Terrett, Nicholas, 187

Wambugu, Florence, 192
Wells, Spencer, 199
Whitson, Peggy, 207
Wüthrich, Kurt, 212

Zewail, Ahmed H., 218